D1415398

Welcome

to MMA's Living Stewardship Series!

As a church-related organization dedicated to helping people integrate their finances with their faith, MMA is pleased to provide this resource as part of the *Living Stewardship Series.*

MMA helps Christians answer God's call to care for and cultivate the gifts God has given them. To accomplish this, we offer products, services, and resources – like this study book. Our goal is to help you understand biblical principles of stewardship while, at the same time, providing real world ways you can incorporate those principles into your everyday living.

The Bible tells us we are to seek wholeness in our lives. In the Gospel of Matthew (5:48) Jesus said in his Sermon on the Mount to *"Be perfect, therefore, as your heavenly Father is perfect."* But, who among us can ever be perfect?

Actually, the Greek word traditionally translated as "perfect" in that verse is *teleios* – which means, "to be whole." *Living Stewardship* is not a series for perfect people, but for people like you who are seeking wholeness. People who don't want to leave their faith in Christ at the church door after Sunday worship. People who want that faith to color how they relate to family and friends, how they work at their jobs, how they spend their money, how they take care of themselves – essentially, how they live.

At MMA, "holistic" refers to the essential interconnectedness of all the elements of Christian stewardship. For the sake of simplicity, we've identified the crucial elements as time, relationships, finances, health, and talents. Integrating all five, and nurturing the relationships between them, produces a healthy life of holistic stewardship.

On the path of your life, you will find the journey easier if you pay attention to all of these areas of your life and recognize how they work together to lead you to the wholeness of God. If one of these elements becomes unbalanced, broken, or disconnected, you experience a lack of wholeness. However, with a strong core (faith) connecting each area, and careful attention to each area as needed, God's love can flow through you and produce wholeness in your life – and in the lives of others you touch.

What is MMA?

MMA helps people and groups integrate their finances with faith values through its expertise in insurance and financial services. Rooted in the Anabaptist faith tradition, MMA also offers practical stewardship education and tools to individuals, congregations, organizations, and businesses.

MMA helps you pursue stewardship solutions through insurance and financial services, charitable giving, and other stewardship resources as well as with our educational resources, such as this study book, and stewardship education events through Stewardship University.

MMA wants to help you integrate your finances and faith and become the best steward of God's resources you can.

This is why we believe *holistic stewardship* involves much more than just the products and services MMA provides. Holistic stewardship looks at the *interconnectedness* that weaves through the areas of our lives. And, as Christians, it's all filtered through our faith in Christ. This faith is what drives the search for wholeness.

How good a steward you are in your finances, can affect your health and your relationships. If you are having trouble with your health, that can affect how you are able to use your talents or your time each day. If you're overcommitted and your day feels too full, you may opt to give short shrift to your children or your job. And on it goes. There are countless ways our search for wholeness is affected by our shortcomings in these areas.

MMA®

Stewardship Solutions

Practical tips to keep you moving!

This study book is on creation care – seemingly not within MMA's mission of helping you integrate your faith and finances. But a closer look, such as provided by this book, reveals that many of the environmental issues we're facing – and many of the questions we're wrestling with concerning the environment – are deeply rooted in money and materialism, topics that are at the core of what MMA helps people think about.

We believe this book will help you look at the environment in a new way. Instead of a technical treatise on global warming, the author begins with a reminder to reawaken our sense of wonder at God's creation. Simply taking the time to view a beautiful sunrise or sunset might be all it takes.

However, because of the holistic nature of stewardship, don't be surprised when we also talk about your talents, time, finances, and relationships – specifically as they relate to creation care.

We'll give you practical ways to implement the suggestions we make here – not just open-ended theories! Each chapter ends with discussion questions you can answer as a group, or individually, that will help you identify areas where you may need to do some repair work.

There's more!

If you like what you learn here, look for other study guides in the *Living Stewardship Series.*

If you want to learn more about us, visit MMA-online, our home on the Web (www.mma-online.org). There you can find more information and tools to help you on your stewardship journey. You'll also find connections to the MMA partners in your area who can help you achieve the steward-ship goals you have for your life.

Creation Care

Keepers of the Earth

by Luke Gascho

To Jane!
May our love
for all that God loves increase.
Luke

MMA®

Stewardship Solutions

Goshen, Indiana

Co-published with Herald Press

Creation Care

Living Stewardship Series

Library of Congress Cataloging-in-Publication Data
Luke Gascho, 1952-
    Creaton care : keepers of the earth / by Luke Gascho
    p. cm.
ISBN 978-0-8361-9467-8 (pbk.)
 1. Ecology–Religious aspects–Christianity. I. Title.
BT695.5.G37 2008
261.8'8–dc22

                                                    2008017563

Cover design by David George & Associates, Inc.
Edited by Michael Ehret

MMA®

Stewardship Solutions

1110 North Main Street
Post Office Box 483
Goshen, IN 46527

Toll-free: (800) 348-7468
Telephone: (574) 533-9511
www.mma-online.org

Acknowledgements

For Becky,

My lifelong companion in caring for creation.

Contents

First Word: A Unique Message

A certain celebrity frog once sang, "It's not easy being green." As we move well into the 21st century, many of us are learning the truth of those lyrics.

- We are concerned about global warming, but we dream of 3,500-square-foot houses with four bedrooms, three bathrooms, and three-car garages – even though our families are smaller than they were 50 years ago.

- We worry about how much we are discarding into landfills, but we keep buying plastic containers and other organizers to hold all our "stuff."

- We are appalled by oil spills that kill wildlife and ruin local ecosystems, but it's hard to give up our beloved gas hogs that get 15 miles per gallon on a good day with a strong wind at our backs.

Any way you look at it, the environment and how we live in it is a complicated topic. One of the major reasons why is the interconnectedness of all things. It's much like a spider web – when you touch a single strand at one edge of the web, it creates movement on other strands on the other side that may at first seem disconnected.

> *The environment and how we live in it is a complicated topic.*

This brings us to one of the major questions you may have. Why is MMA, a stewardship organization focused on the integration of faith and finances, sponsoring a book about the environment? Simple. Many of the issues we're faced with – and questions we have about – the environment are deeply rooted in money and materialism, topics that are at the core of what MMA helps people think about.

We believe this book will help you look at the environment in a new way. Instead of a technical treatise on global warming, the author begins with a reminder to reawaken our sense of wonder at God's creation. Simply taking the time to view a beautiful sunrise or sunset might be all it takes.

The issues and questions are deeply rooted in money and materialism.

In subsequent chapters, you will look at the biblical case for being concerned about the environment along with the ethics and justice issues involved. Another significant issue addressed is the economics of creation care.

It is only after all that background that you will begin delving into what most people would expect in a study on the environment – water quality, energy use, and climate change. Even here, however, the issues may be framed a bit differently than you have experienced before.

Finally, you will be asked to think about creation care in terms beyond yourself. What does creation care mean for your congregation – and for wider denominational groups?

At the end of the day, we hope you come to a fuller understanding of the intricate connections in God's creation. Solving global warming may not be as simple as shutting down all coal-fired generating plants. Just like the spider web, we need to consider what happens to all the other threads when we decide to move even one.

Being green may not be easy – but perhaps you'll find it's what you want to be, too.

Stewardship and Wonder

1

The best remedy for those who are afraid, lonely or unhappy is to go outside, somewhere where they can be quiet, alone with the heavens, nature and God. Because only then does one feel that all is as it should be and that God wishes to see people happy, amidst the simple beauty of nature. As long as this exists, and it certainly always will, I know that then there will always be comfort for every sorrow, whatever the circumstances may be. And I firmly believe that nature brings solace in all troubles. — Anne Frank, February 23, 1944

In awe and wonder

Examining the wings of a monarch butterfly causes my mind to be filled with wonder. Here is a small creature that can soar gently in the breezes around my garden, but also one that can migrate thousands of miles to southern Mexico. The delicate beauty of the wings captures my attention during those brief moments when the butterfly lands on a flower. The tiny scales that make up its wings carry a pattern that is unique to all monarchs. These wings were once folded up in a chrysalis. It was the slow pumping of a liquid through the veins of the wings that has given them the aerodynamic strength needed to fly.

My thoughts, as well as yours, can go from one marvelous creature to another – some are exquisitely small and others profoundly large. The movements of the creatures are fascinating as they vary from slow to fast, darting to clinging, slithering to gliding, galloping to creeping, and diving to leaping.

And then there is the beauty in flower patterns, shapes and colors. I enjoy early morning walks through the native plantings in my yard during the summer. Here I find the delicate blues of spiderwort and blue false indigo, the vibrant orange and yellow of butterfly weed, the purple of ironweed, and the broad green leaves of the prairie dock.

Our minds can go deep and wide throughout the universe we have encountered. The vistas that we drink in with our eyes can vary, too. For some of us it is the landscape of forested wetlands, for others it is the prairie, or the desert, or the mountains or the ocean view or the gentle rolling flint hills. In each case we are drawn with awe and wonder to the complexity of the landscape with its colors, rhythmic shapes and flowing expanse.

Our favorite views and settings bring peace and belonging to our inner being. There are messages that we experience by spending time in God's creation – messages that are missed if we don't give time to interacting with this natural beauty. The psalmist understood the meaning of wonder when he wrote:

> The heavens are telling the glory of God;
>   and the firmament proclaims his handiwork.
> Day to day pours forth speech,
>   and night to night declares knowledge.
> There is no speech, nor are there words;
>   their voice is not heard;
> yet their voice goes out through all the earth,
>   and their words to the end of the world. – Psalm 19:1-4

Knowing these messages is important to our awareness and alertness to God's action that invigorates the cosmos. It is an action of love that is the source of all life as seen in the handiwork of God. The messages found in the beauty and wonder of the world are integral to our fuller understanding of the incarnation of Christ. God chose to have his son take on flesh and experience this world from the complete human perspective. The incarnation assists us in knowing the answer to the question asked by the psalmist when he observes the beauty of the night sky while watching his sheep,

> When I look at your heavens, the work of your fingers,
>   the moon and the stars that you have established;
> what are human beings that you are mindful of them,
>   mortals that you care for them? – Psalm 8:3-4

We are given a relationship with God and with all things that he created. Even though we can appear to be insignificant in comparison with other parts of the universe, we have a relational role as human beings with the rest of creation.

Throughout the scripture we find frequent incidences where God and people interact in a place filled with natural wonder. Consider the following examples.

1. Adam and Eve were placed in the Garden of Eden to till and to keep it. It was here that they walked and talked with God. Imagine the conversations! It had to be amazing for them to ponder with God's input the beauty of each item in the garden.

2. Job had a life full of blessing, but one that turned to suffering in the middle of suffering. There was little consolation found in the many words of his friends. But then God spoke out of the whirlwind (Job 38:1). The four chapters that follow are filled with God using multiple examples from nature to point out the care and relationship that God has with all things – including Job.

3. Moses' life is interspersed with times of meeting God in the great outdoors. He experiences the desert, the bush on fire, the moving of waters, and the treks up the mountain all as part of knowing who he was and how to faithfully lead God's people.

4. The Psalmist repeatedly describes encounters with God in nature. Psalm 104 poignantly illustrates the relationship between nature, human beings, and God. All of the illustrations by the poet are resplendent with awe and wonder. In everything the Creator receives the praise.

5. The prophets often meet God in the places remote from human habitation. It is in these settings that they hear the voice of God. This is vividly recounted in the story of Elijah as he meets God on the mountain. While fleeing for his life, Elijah encounters wind, earthquake, and fire. It is then in the "sound of sheer silence" that he hears the voice of God.

6. Jesus delighted in speaking to his followers and the crowds by the sea, on the mountains and in the fields. It was here that he would draw the listeners' attention to created things in order to make his point. A great example of his teaching with nature is when he said, "consider the lilies of the field."

Admiration, observation and wonder of creation are spiritual acts. As Christians we benefit by seeking times to experience the fullness of God's creation. Nature guides – people who are trained in ecology – lead us in our understanding just as teachers and pastors – people trained in theology – are able to open texts of scripture with fresh insights. But it is also important to learn to step out on our own and take walks with God in the simplicity and the grandness of creation – and be filled with awe and wonder.

Group questions

1. *What are your favorite creatures, plants, and vistas that bring delight to your soul? (Recall good memories from playing outside as a child.)*

2. *What is it that draws you to this admiration and wonder of created things? How can our experiences of wonder in the created world enhance our worship of God – the Creator?*

3. *How do you interpret the meaning of God's relationship to us as described in Psalm 8:3-4?*

4. *Consider and list the connections between the wonder that we see in the Earth around us and the glory that God reveals to us.*

The Earth: Our Home

Our dwelling place is a world we did not make. Oceans of air surround us. We stand on land consisting of a relatively thin coat of soil laid over bedrock. Pools of water of all sizes are found intersecting with the land, as well as in aquifers beneath our feet. Around us are many species of plant that are adapted to the types of soil and the amount of available water. Creatures both visible and invisible inhabit the landscape and the seascape.

The Earth rotates slowly and consistently, bringing us day and night. During the day we see because of the rays of the sun – and its energy rejuvenates all systems. At night we marvel at the expanse of the universe as marked out by the moon and stars. The tilt of the Earth combined with our movement around the sun brings the changing seasons. "As long as the earth endures, seedtime and harvest, cold and heat, summer and winter, day and night, shall not cease." (Gen. 8:22)

Who are we anyway in the midst of this location and story? Are we not creatures from the same source – the God of the universe? What is the role we are called to within the Earth and its systems? Are not our roles as diverse as the beauty we see around us? Part of the answer to these questions is found as we continue reading in Psalm 8:5-6.

> "Yet you have made them a little lower than God,
>   and crowned them with glory and honor.
> You have given them dominion over the works of your hands;
>   you have put all things under their feet."

We are modeled after God – in God's image. This means that God's way of relating to the creation is a design for our roles of relating to the Earth and its systems. We are not in the same position as God, but we have been given a special place within the created order. Knowing that place and how to act within our stewardship role are central issues of this study. It is important to notice that the psalmist uses the terminology of "glory and honor" to describe our position. This is not a position of loftiness and 'a lording it over' all other things within creation. Rather the terms characterize the attitudes we should have as we relate to creation. Glory implies rejoicing and marveling over the beauty in nature. Honor signals the importance of integrity and respect.

Understanding this stated position and the attitudes that go along with it adds clarity to a challenging word found in verse 6 – dominion. This term is often seen as a license for humans to do whatever they want with the Earth and its resources. Many times in our history domination was our mode of operation. Consumption and destruction are the tragic results. If we look at dominion more closely, we can see instead the concept of a compassionate and benevolent king.

This king – modeled after God – is one who desires healthy relationships with everything in the kingdom. By following this design, glory and honor are experienced by all members of the created order. Unfortunately, most human kings have not demonstrated this kind of relationship with the Earth – its inhabitants and its resources.

In seeking terms to help us know the role we should play as Christians in relationship to creation, consider the Old English word, stigweard. At first glance, it is hard to see the relevance of this word in this study, but by looking at it in two parts we make an important discovery. 'Stig' means hall, house, or pen. For example we know the term pigsty, which is the pen where a pig lives. The portion 'weard' carries the concept of guard or keeper. The combination of the two parts of stigweard translates into "keeper of the house." It is to be noted that this old word – stigweard – is the basis for our word steward. Being keepers of the house – our home, the Earth – is the role that God has called us to perform. We could just as easily say, "We are to be earthkeepers."

A word that is used frequently in environmental studies is the term ecology. The origin of this word's prefix 'eco' comes from the Greek word for home or household. When the suffix 'ology' is applied to the prefix 'eco,' we have the basic meaning of ecology. It is the study of our home. Not only is it the study of the human in relationship to our surrounding environment, it is also the study of the habitats and interconnections of all other organisms with their settings or homes.

In examining stewardship and ecology side by side, we discover that one term describes our role of keeping and caring for the Earth and the other notes the importance of learning to know the relationship between all things. Ecology informs stewardship and stewardship provides balance in ecological health. These are the kinds of roles we are called to explore as Christians in this study of the stewardship of the environment. An Anabaptist belief statement on Christian stewardship clearly states this role.

> "As servants of God, our primary vocation is to be stewards of God's household."[1]

Exploring the interconnectedness of all things is essential to becoming well-informed stewards. Unfortunately in our western civilization, we are disconnected from a more intimate relationship and awareness of the creation and its intricate systems. We live in houses that are conditioned to keep us from extreme fluctuation of temperature. We travel in conditioned cars to arrive at work places that are conditioned as well. As we move through life from conditioned box to conditioned box, we end up isolated from much of God's creation – the very organisms he called us to relate to. In earlier eras, our ancestors knew the relationship better because they were in daily contact with the multiple aspects of the environment. Awareness of this disconnect can be the starting point for us to reengage in the relational design God models for us.

If we spend time in nature and experience awe and wonder, we can be revitalized in our call to be God's stewards of all that he has given to us. It is in seeing the beauty around us that we are reminded that the Earth is a gift to us. We also see the parts of creation that have been and are being broken as a result of our human carelessness and selfishness.

The wonder we find in nature is a valuable inspiration for our lives and for care of the Earth. We care for what we love and what we have been given. This is a principle that should inform our acts of stewardship. It should also guide our acts of creation care.

God formed and created. God's spirit moved across the face of the Earth. God breathed and gave life. God walked and talked with Adam and Eve in the Garden of Eden. This was spiritual relationship at its best. May we seek the same.

Prayer:

God, we praise you for the beauty and wonder of the created world.

We thank you for giving us the gift of life.

Thank you for senses with which we can observe your handiwork

and for minds to marvel in your grandeur.

We acknowledge the brokenness that we have caused

when we are not good keepers of the house.

Make us mindful of the important role of being stewards of all things.

You have blessed us with the gift of creation.

We will explore it joyfully.

May we find wonder.

May we seek right relationships.

May we care as you care.

Amen.

End questions

1. *How does the "keeper of the house" concept add to your understanding of your role in biblical, holistic stewardship?*

2. *What difference does it make whether we see the Earth's resources as things to be used up or as resources to be cared for?*

3. *What are the ways in which you see the wonder in creation informing your acts of stewardship related to all God has entrusted to us?*

4. *How might we best interpret God's relationship with all things as a model for our care of the Earth?*

An activity to do at home:

> *Go outside with your camera with the intent to focus on a wonder in nature. Allow this time to be a double focusing – one of your camera, the other of your spiritual being.*

Sources

1   "Article 21: Christian Stewardship," in Confession of Faith in a Mennonite Perspective (Scottdale, Pennsylvania: Herald Press, 1995), 77.

Finding My Place

2

"… a ditch somewhere – or a creek, meadow, woodlot, or marsh … These are places of initiation, where the borders between ourselves and other creatures break down, where the earth gets under our nails and a sense of place gets under our skin. Everybody has a ditch, or ought to. For only the ditches and the fields, the woods, the ravines – can teach us to care enough for all the land." – Robert Michael Pyle, The Thunder Tree, 1993

Take a walk

I love the land and lakes of northern Minnesota – the places of my childhood. There I experienced the four seasons with all their wonder and extremes. The woods, meadows, and fields of our small dairy farm provided a place to roam and discover. The seasonal stream that ran through the pasture was a great place to slosh and romp. A short walk from the farm took me to a large lake where I could fish, swim, skate, float on a raft, and later explore by boat. The dense woods had their own aura that included fear of what might be lurking – and the delight of seeing wildlife.

This place included the role of work in addition to the sense of wonder. The farm was a setting to do chores in the rhythm of the seasons. Here I learned the cycles of life and death, planting and harvesting, success and failure. Sustaining life meant developing the art and skill of growing food for our table and for our livestock. It was essential to find ways to work with nature and the seasons rather than trying to overpower the harsh conditions. Much of the work could only be accomplished by the joint efforts of neighbors. Appreciation for the paradoxes of beauty and sweat, bounty and limit, splendor and labor shaped my understanding of my relationship to the created world. This was my home.

Our stories of place are important in our development and our identity.

Each person has a different experience of encounter with place. Our stories of place are important in our development and our identity. Again from my personal vantage point, I enjoyed interacting with my children with the western Pennsylvania setting where they grew up. We had walks in the woods and on the ridges of the Laurel Highlands. There were the times of digging, jumping, swinging, chopping, and hiding in all kinds of locations in the hollow where we lived.

> "And there was a garden that was a central part of providing the food for our table – including the labor of raising and preserving that food. We also experienced a major shift in location during the year we lived in San Juan, Puerto Rico. It took an intentional regrouping within our family, which led to enjoying our place in the urban setting. And now we can watch as our children find their sense of place in various locations scattered across the country. From the beginning of the earth, God interacted with place in a personal way. The story of the creation of humans includes God situating Adam and Eve in a unique place. Genesis 2:8 says, "And the LORD God planted a garden in Eden, in the east; and there he put the man whom he had formed." This was where Adam and Eve walked and communed with God, learned about all the created things, gained understanding about who they were, and also experienced a break in their relationship with God. The Garden of Eden represents our past and beckons us with a reminder of the future. Knowing our place and where we are is important in knowing who we are. It is essential that we gain insights into our place, which is where God has put us. Eugene Peterson develops this concept in his book *Christ Plays in Ten Thousand Places*. Everything that the Creator God does in forming us humans is done in place. It follows from this that since we are his creatures and can hardly escape the conditions of our making, for us everything that has to do with God is also in place. All living is local: this land, this neighborhood, these trees and streets and houses, this work, these people."[1]

We do not exist in a void. We have been given a place within a community made up of all aspects of nature, including people. To not know this place leaves us with a lack, an emptiness – of which we may not even be aware. Completeness is enhanced with finding and growing in relation to our place. There is a sacredness to be recognized in our sense of place. As Wendell Berry says, "I have learned to see my native landscape and neighborhood as a place unique in the world, a work of God, possessed of an inherent sanctity that mocks any human valuation that can be put upon it."[2]

There is a sacredness to be recognized in our sense of place.

Our way of life in North America leads us away from a sense of wonder and place. The mechanical and technological aspects of our lifestyle make it easy to operate in a realm separate from nature. We enjoy safety and comforts not available two generations ago. Visual and auditory inputs that are electronically produced stimulate our minds. This has become our norm, our way of functioning. This represents a loss for many adults, but may have even greater impact on children. The influence of this change, moving from an engagement with nature to a television- and computer-stimulated culture, is explored in the book *Last Child in the Woods: Saving Our Children from Nature-Deficit Disorder,* by Richard Louv. Louv writes about how unstructured activity has almost disappeared for many American children. He persuasively states that children's total well-being is at stake. Based on research, he develops the concept of nature-deficit disorder and describes what happens when children are disconnected from direct nature experiences. By weighing the consequences of the disorder, we also can become more aware of how blessed our children can be – biologically, cognitively, and spiritually – through positive physical connection to nature. Indeed, the new research focuses not so much on what is lost when nature fades, but on what is gained in the presence of the natural world.[3]

Contact with nature lessens stress and heightens children's creativity and concentration. Above all, it increases their joy in life. The same is true for adults. We need to seek restorative activities that bring us back into regular interaction with created things. This can be done in various ways whether we are in urban or rural settings. Bird-watching, walks to note the changing seasons, sitting in quietness, planting a garden, standing by water, or gazing into the sky are all opportunities for reconnection with where God has placed us.

We need to seek restorative activities that bring us back into regular interaction with created things.

Our memories of place also help us find a sense of home within our minds. Some of us can recall the place we were born because we lived there throughout our childhood. Others have memories of places where they grew up that are quite detached from the place of their birth due to geographical changes for our families. Still others moved with great frequency due to changing dynamics and jobs. In each case we can recall the settings that made us feel contented and connected.

Group questions

1. *What places in nature gave you the greatest sense of comfort, peace, or stillness at an earlier time in your life?*

2. *What is a place in creation (nature) where you feel "at home" (a sense of belonging, comfort, or rest) today?*

3. *What are the ways you have experienced a disconnection from creation or nature? In what ways are you strengthened by time in nature?*

A place to call home

Belonging is a central desire of all people. This concept is nurtured in the places we call home. In the context of our study, we are acknowledging that the Earth is our present home. It is where we have been placed. We have already noted both the value of place that comes through a connection to it and that we live in a time when our culture threatens to disconnect us from a healthy relationship with nature. To further develop our understanding of the value of place, we will examine six key principles of place.

The Earth is our present home.

Place reflects God's relationship to all things

The Creator God has given us the Earth as our home. This place is full of wonderful designs and patterns, which reflect the image of God. We have been made in that image, giving us a unique position of relationship to all things. The relationship spans time in multiple dimensions. The everlasting relationship with God is described in a song of Moses as recorded in Psalm 90.

> "Lord, you have been our dwelling place in all generations.
> Before the mountains were brought forth,
> or ever you had formed the earth and the world,
> from everlasting to everlasting you are God." (Psalm 90:1-2)

As people, we have been in a dwelling relationship with God throughout all time. God seeks that relationship with us. The ability to find comfort in place is because of God's relationship to all of creation. We experience God within this inclusive relationship.

Place provides a model for connectedness

We live in an intricate web of a world in which all creatures are interrelated. The study of ecology is fascinating because of these relationships. As we observe nature – its individual components, the way organisms depend on each other, the changes that occur within ecosystems – we are reminded that we too are part of this living dynamic. On the one hand, it can make us feel insignificant, but on the other hand it reminds us of the need for relationships.

It is important that we do not forget about the relationship of all people within this concept of connectedness. We function as human beings because of the sustaining role of the rest of creation – and the fact that Christ sustains all things (Colossians 1:17). Our interconnectedness means getting to know our neighbors. Just like the interdependence in nature, we are healthier when there are strong relationships within our communities.

Place enhances our sense of wonder

The created world is full of a greater complexity than we can possibly imagine. The diversity of plant and animal species is a marvel to behold. Each creature reveals a splendor that fills our minds with wonder. The colors, patterns, and uniqueness catch our attention – and remind us there is a place and purpose for all.

If we discard an object … it remains there. There is no "away" when we throw things away.

This view into the wonder found in each place we call home also shows us the delicate balance between all things. We recognize how easily the systems and relationships can be disrupted. If we discard an object such as an aluminum can, it remains there. There is no "away" when we throw things away. There is a need for conservation in order to maintain wholeness.

Place is more than a material resource

It is easy to objectify nature and simply turn it into a list of resources for us to use. Using materials from our surroundings for food and shelter is part of the relationship we have been given. But it becomes destructive when we move into a consumptive mode that promotes "using up" nature without thought to replenishment. We live in a world of limits, which is a concept we have ignored for too long. More is not always available.

One of the shortfalls of our western worldview is that we tend to see humans as separate from nature rather than a part of it. The use of the term "environment" also implies this understanding. Environment is what is around us. This concept causes us to be detached. I prefer to use the word creation rather than environment. Creation includes all in God's plan. It means we are part of the place – not just in the place.

Place promotes generational responsibility

We have been given a wonderful heritage within creation. As the generations pass, the heritage of place is passed on. David states this so eloquently in Psalm 16:6, "The lines have fallen to me in pleasant places; Indeed, my heritage is beautiful to me."

Treating the Earth in such a regenerative way will grant future generations a home that is also a pleasant place and a good heritage. Our actions have a ripple effect that extends across the globe in the present and on to future generations. Steve Bouma-Prediger emphasizes this concept: "If we wish to properly care for our homes — not only for ourselves but for our children and our children's children – then we, and all our fellow dwellers in our place, must love our homes."[4]

Place is the setting for rest

Learning to know our place helps us appreciate the need for rest and renewal. The creation has a high level of resilience, but if rest is not given, it will wither. The Earth's systems are dynamic. Living organisms are constantly going through renewal and change. If the natural cycles were not dynamic, changing, and adaptive they would cease to exist.

This pattern seen in place is an essential model for humans to participate in as well. We, too, need to experience Sabbath — and we need to extend it to all relationships.

Caring for our place

All people — and especially people of faith — have an important role to play in relation to place. It is the role of caring. Our connectedness leads us to wholeness and a desire to care. The act of stewarding is one of caring and keeping the home. This responsibility was succinctly given to Adam and Eve in Genesis 2:15: "The LORD God took the man and put him in the garden of Eden to till it and keep it."

We need to grow in loving all that God loves.

We tend to think of the land serving us and supplying what we need. God instead points to a better relationship of being a servant to the land. Additionally, we are to "keep" or protect the land. Creation is something that can be destroyed or degraded. Protecting the land includes the rest and renewal that is essential for wholeness.

Right relationship is a key in good stewardship of creation. We need to grow in loving all that God loves. Steve Bouma-Prediger summarizes the focus of this lesson well by saying,

> We care for only what we love. We love only what we know. We truly know only what we experience. If we do not know our place — know it in more than a passing, cursory way, know it intimately and personally — then we are destined to use and abuse it. For we will care for our home place only if we love it, and we will love it only if we know it, and we will know it only if we experience it firsthand.[5]

Prayer:

Lord of Creation, thank you for setting us in a pleasant place,

and for giving us a godly heritage.

We see your beauty and wonder displayed all around us.

You beckon us to come, enjoy, and participate in right relationship.

We confess we have been influenced by the gods of this age,

and have withdrawn from our interconnectedness.

The result is that we have caused harm to creation.

We commit to renewing ourselves to a holistic sense of place.

May we contribute to the health of creation for the coming generations.

Amen.

End questions

1. *What are your hopes for future generations in terms of a healthy place to live?*

2. *How do you experience rest in your favorite place? How do you extend Sabbath to yourself and others?*

3. *In what ways can you practice the concept of bringing "rest and renewal to the land and everything that lives on it?"*

An activity to do at home:

> *Spend a half hour in a favorite place in nature. Note the things that are very familiar to you, as well as those that have changed. What do you hear from God during this time in sacred space?*

Sources

1   Eugene Peterson, *Christ Plays in Ten Thousand Places* (Grand Rapids: Eerdmans, 2005), 72.

2   Wendell Berry, *The Way of Ignorance: And other Essays*, <u>Imagination in Place</u> (Emeryville, Calif.: Shoemaker and Hoard, 2006), 51.

3   Richard Louv, *Last Child in the Woods: Saving our Children from Nature-Deficit Disorder* (Chapel Hill, N.C.: Algonquin Books, 2006), 34-35.

4   Steven Bouma-Prediger, *For the Beauty of the Earth: A Christian Vision for Creation Care* (Grand Rapids: Baker Academic, 2001), 38.

5   Steven Bouma-Prediger, *For the Beauty of the Earth: A Christian Vision for Creation Care* (Grand Rapids: Baker Academic, 2001), 37.

Creation Care Theology

"*Our predicament now, I believe, requires us to learn to read and understand the Bible in the light of the present fact of Creation. This would seem to be a requirement both for Christians and for everyone concerned, but it entails a long work of true criticism that is, of careful and judicious study, not dismissal. It entails, furthermore, the making of very precise distinctions between biblical instruction and the behavior of those peoples supposed to have been biblically instructed.*" — Wendell Berry[1]

Walking with God

It was a marvelous fall day as I walked along the edge of the Elkhart River. The water flowed slowly, providing wonderful reflections of the fall colors in the trees and the deep blue of the sky. My hike through the forested flood plain was a time of rejuvenation.

The low-lying area through which I walked makes up much of the 63 acres of land owned by Waterford Mennonite Church, my home congregation. I love this sanctuary. I join many others in worshiping in the sanctuary within the church building on Sunday mornings, but what an inspiration to head out into our other sanctuary, too.

> *The scenes I experience as I walk along the riverbank reflect God's act of love.*

My walks along the shoreline offer a place and a time to reflect on our Christian understandings and beliefs about God and the universe. The aesthetics of this sanctuary are not to be matched by human architecture. There is grandeur in the architecture of nature that supersedes the human capacity for making things fit together and display beauty. In this setting, one can observe how systems work and ask questions about God's involvement in what is seen. King Solomon spent much time reflecting and writing on the wisdom of God and nature was often the context for his thinking.

> "When he established the heavens, I was there,
>   when he drew a circle on the face of the deep,
> when he made firm the skies above,
>   when he established the fountains of the deep,
> when he assigned to the sea its limit,
>   so that the waters might not transgress his command,
> when he marked out the foundations of the earth,
> then I was beside him, like a master worker;
>   and I was daily his delight,
>   rejoicing before him always,
> rejoicing in his inhabited world
>   and delighting in the human race." (Proverbs 8:27-31)

The walk along the Elkhart River typifies the wise thoughts of Solomon. Like him, I saw the sun reflected in the water, the dome of the sky, the bottom of the river, the shore to limit the water – and all of this on a firm foundation. I experienced the grace of what God has given to us. I saw the human impact on this setting – the silting from soil erosion, discarded bottles, spots of oil floating on the water, an abandoned lawn chair, and invasive plant species. A host of questions mixed within my thinking. Why is the universe here? What is God's relationship with creation? What does God expect our role to be in relation to the Earth?

First, let's consider why we believe the universe is here. The focus of this study is not about the question of how the Earth came to be. The controversies and tensions over the origins of the universe can distract us from focusing on the need for us to be good stewards of the Earth. I suggest conversations on origins should be saved for another time and place. We believe that love is a central characteristic of God and this love is utterly self-giving. The statement from the *Confession of Faith in a Mennonite Perspective* articulates this belief. "We believe that the universe has been called into being as an expression of God's love and sovereign freedom alone.[2]

The scenes I experience as I walk along the riverbank reflect God's act of love (see John 1:1-4). The brilliant points of light I see in the night sky show the love of God in the universe (see Psalm 19). The universe is an expression of God, which is quite different from it being an extension of God. The agape love of God is on display in the world. John 3:16 says, "For God so loved the world ..." The word "world" means cosmos or universe. The world is everything that the triune God has made – and it is all an expression of who God is, which is love.

> *The agape love of God is on display in the world.*

Second, what do we understand as God's relationship with creation? From the beginning of Scripture in Genesis 1 to the book of Revelation, we find God in personal relation with the universe. God is not a distant being who looks on the Earth with detachment. In Genesis 1 we find God expressing delight in all he has made. Each part of creation is called good. This delight is possible only when there is relationship. In Revelation 21:3 a voice says, "See the home of God is among mortals. He will dwell with them ..." God continues this close relational role throughout all history.

There are differing views as to God's intersection with our space – the Earth. Some subscribe to the idea God is everything and everything is God, which is pantheism. Another view is panentheism where it is understood that everything exists within God. Others, such as Deists, see God as distant and one who has left the world to run on its own. Questions remain about interrelationships in each of these perspectives. What, then, is a Christian view?

An excellent way of answering this grows out of bringing everything under the lordship of Jesus Christ. There is a strong relational position that emerges out of the radical and living nature of redemption through Christ. Thomas Finger explains the God-world relationship:

> To affirm that God's being is distinct from creation is, as we have seen, by no means to affirm that God is distant, as panentheists often suppose. If God takes the sufferings of creatures and indwells them out of grace, God comes astoundingly close to them, and God's love for them is greater than would be if they were naturally interconnected with God.[3]

The triune God's love for all of creation and people is recorded in the great poem of Isaiah 40. Comfort is extended to a suffering people through reminders of God's presence within the world. Verse 21 is part of a series of many questions.

> "Do you not know? Have you not heard?
> Has it not been declared to you from the beginning?
> Have you not understood from the foundations of the earth?"

The answer to each question is a reminder of God's interactive nature with people and all of creation. Grace is extended as a sign of this relationship – the grace ultimately found in Christ. God's relationship is constant and reassuring – it does not faint or grow weary.

Group questions

1. Share your experiences in sanctuaries not made with human hands.

2. How have you observed God's love by experiences in nature? How are these observations the same and/or different from experiences you have with people?

3. What are your personal and biblical insights that describe God's relational qualities with:

   a. People?

   b. Plants?

   c. Animals?

   d. Ecosystems?

   e. The universe?

Act in God's image

My walk along the Elkhart River revealed signs of beauty and brokenness. As I noted earlier, I saw the evidence of human impact that did not add to the health of the river system. This raises my third question. What does our theological understanding say about our roles in relating to creation? Consider the early story from Genesis where we hear about God's view of humans and their roles.

> "So God created humankind in his image, in the image of God he created them; male and female he created them. God blessed them, and God said to them, 'Be fruitful and multiply, and fill the earth and subdue it; and have dominion over the fish of the sea and over the birds of the air and over every living thing that moves upon the earth'." (Genesis 1:27-28)

The first observation to make is that we have been created in God's image. By coupling this understanding with our earlier notation that creation is an expression of God's love, it is clear that loving and caring for creation is part of how we are to faithfully display God's image. We are to do as he does.

What does our theological understanding say about our roles in relating to creation?

The next set of concepts – to fill, to subdue, and to have dominion – in this passage can be problematic in helping us understand our role within the Earth's systems. The filling of the Earth challenges the carrying capacity of the planet. There are limits to what the Earth's systems can support. If any species populates the Earth beyond the capacity of the Earth's systems to support life, then this becomes a destructive act and is contradictory to the love God has for the world. In light of the rest of the Bible, we are to be faithful in replenishing without causing degradation.

Subdue is a command that is often interpreted as "lording over" other things. But, there are other meanings more in keeping with the context, such as "to arrange or order" and "to name." Adam was faithful to this command when he named all the animals. This act of ordering the animals with a set of names helped form appropriate relationships between humans and the rest of creation. If we know the name of something, we are more apt to care for it – person, plant, rock, or animal.

Dominion is another term that can be misapplied as domination. In its best sense, dominion refers to the benevolent king who cares compassionately for all in his domain. This kind of relationship encourages health, a flourishing, and peace in the kingdom (see Psalm 72). In its worst sense, dominion can become a dominating force that denies rights and causes disruptive behaviors. I believe each of the three terms can guide us in right relationships and provide a good basis for creation care. We are called to a careful discernment of these instructions and to understand them in light of the full context of the Scripture. As previously noted, God commands us to "till and keep" or "serve and protect" his creation. (Genesis 2:15)

Be reconcilers

The Old Testament gives us redemption stories played out in the context of God's relationship with creation. Looking to the New Testament gives us added understanding of God's view on our role of caring for the Earth. Redemption through Christ is the center point of faith. "God has made provision for the salvation of humanity and the redemption of creation."[4]

We, as ones who have been reconciled, are also to be reconcilers through our words and practice.

The reconciling pattern of the triune God at work in the world is communicated in an early Christian hymn found in Colossians 1:15-20. Consider this passage and the way it relates to "all creation." This hymn states the creative, sustaining, and reconciling acts of Christ are for all of creation, which includes humankind. The message of 2 Corinthians 5:19 ties together a number of themes that relate to redemption and reconciliation: "...that is, in Christ God was reconciling the world to himself, not counting their trespasses against them, and entrusting the message of reconciliation to us."

Reconciliation is extended to humans through Christ. The world – or cosmos – is also included in this reconciliation process. And we, as ones who have been reconciled, are to be reconcilers through our words and practice. Dorothy Jean Weaver adds insights to the message Christ brings to us:

> God the Sovereign Ruler is by the same token God the Creator. And Christ the Redeemer is none other than Christ the Agent of Creation. Accordingly, to acknowledge God's rule and God's redemption is to stand in reverence before God the Creator, to submit our lives to the claims of Christ the Agent of Creation, and to treat with utmost care and respect all that which God has brought into being through Christ.[5]

We have experienced God's grace through Christ. How privileged we are to be messengers of this good news, which includes the acts of being reconcilers to all of creation! Grace is to be our theme as earthkeepers.

Anticipate the future

Every time we pray the Lord's Prayer we make a statement about our hope for the Earth today and into the future. We repeat from Matthew 6:10, "Your kingdom come. Your will be done, on earth as it is in heaven." We are asking for the renewal of all things on Earth as it is represented in heaven. James Jones calls this "the earthing of heaven."[6] God's eternal purposes, as shown through the resurrection, include the physical as well as the spiritual.

Our view of the future influences the way we choose to care for creation. In Revelation 21:1 we read of a "new heaven and a new earth." For this study we will work with just two perspectives on this passage. One view is that the current Earth is destroyed by fire. A text that is used to support this understanding is 2 Peter 3:10-13. While many translations include a description of fire reducing the Earth to nothing, the fire can also appropriately be seen as a purification and preparation for the new to come.

Some Christians take the passages that predict a destruction of the Earth as a basis for not being involved in work related to our environment. They believe God has given the Earth to us as a resource to be used and that ecological stewardship has little value since the Earth will be destroyed when Christ returns. Heaven is a separate place with no physical connection to the present Earth.

> *Our view of the future influences the way we choose to care for creation.*

The second view is that in the future there will be a full restoration of the Earth, which is part of the new heaven and new Earth. In Revelation 21:2, John sees the New Jerusalem coming down to Earth, which represents the renewal of the Earth. Romans 8:18-25 speaks of the groaning of creation and the setting free of this bondage. Here we also find that redemption and the resurrection of our bodies includes the physical and the spiritual. Thomas Finger expands this point:

> In general, the redemptive pattern throughout Scripture speaks not so much of humans going to heaven as of heaven coming ever more fully to Earth. It

is appropriate to image, with the last chapters of Revelation, the New Heaven and New Earth not as wholly disconnected from the present Earth, but as coming down upon the latter and transforming it through God's presence. This means that whatever we do to preserve and enhance our environment will in some way be preserved and transformed in the final state.[7]

The lack of ecological care Christians extend to the Earth baffles many unbelievers.

In fact many people accuse Christians of being the cause of the degradation of the Earth because of the way various texts have been interpreted and practiced. We have an opportunity to witness to the love of God by reexamining our theology of creation care and taking steps toward a holistic stewardship of the Earth.

Prayer:

Oh Lord, Maker of Heaven and Earth,

We thank you for designing sanctuaries in creation for us to enjoy.

We worship you because of your creative acts.

This world is truly an expression of your love.

We confess we often struggle with understanding the

messages you intend to be filled with hope.

We seek the wisdom of your Spirit to give understanding and clarity

so that our actions may reflect your fullness.

May the grace you have extended to us

be reflected in the care we give to all things.

Amen.

End questions

1. *The beauty we see in nature's sanctuaries captivates us. On the other hand, our actions toward the Earth can be very consumptive. What are examples of this tension in your life?*

2. *Examine Genesis 1-2. Make a list of the ways our relationships with God and all creation are described in these passages. What does this teach us about caring for creation?*

3. *Interact with the concept of "the earthing of heaven." How do you agree and/or disagree with the views on the future of the Earth? What connections do you draw from these views that might inform your care of the Earth?*

An activity to do at home:

> *Pray through the Lord's Prayer slowly several times. Reflect on the ways in which this prayer represents ecological wholeness.*

Sources

1   Wendell Berry, "Christianity and the Survival of Creation," in *Sex, Economy, Freedom & Community* (New York: Pantheon Books, 1992), 94-95.

2   "Article 5: Creation and Divine Providence," in *Confession of Faith in a Mennonite Perspective* (Scottdale, Pa.: Herald Press, 1995), 25.

3   Thomas Finger, "An Anabaptist/Mennonite Theology of Creation," in *Creation and the Environment: An Anabaptist Perspective on a Sustainable World*, ed. Calvin Redekop (Baltimore: Johns Hopkins University Press, 2000), 163.

4   "Article 6: The Creation and Calling of Human Beings," in *Confession of Faith in a Mennonite Perspective* (Scottdale, Pa.: Herald Press, 1995), 28.

5   Dorothy Jean Weaver, "The New Testament and the Environment," in *Creation and the Environment: An Anabaptist Perspective on a Sustainable World*, ed. Calvin Redekop (Baltimore: Johns Hopkins University Press, 2000), 124-125.

6   James Jones, *Jesus and the Earth* (London: Society for Promoting Christian Knowledge, 2003), 27.

7   Thomas Finger, "An Anabaptist/Mennonite Theology of Creation," in *Creation and the Environment: An Anabaptist Perspective on a Sustainable World*, ed. Calvin Redekop (Baltimore: Johns Hopkins University Press, 2000), 168-169.

Creation Care Ethics

"To desecrate the earth and despoil the soil is not just a crime against humanity; it is blasphemy, for it is to undo the creative and redemptive work of God in Christ. All things came into being not for us but for him." —James Jones[1]

A snake on the path

Gazing out across the Laurel Mountains is good for the soul. It was a much-loved view for our family when we lived in western Pennsylvania. Some of our best family memories were the times of hiking and camping on the Laurel Highlands Trail that followed the tops of the mountain ridges. One summer, my two sons and I decided we would take several days to hike a third of the trail by starting at the southern trailhead. The first day involved several miles that were all uphill. As we tired in the heat, our heads began to droop to the point where we just watched our feet take one step at a time. "When will we get to the top?" was the thought on our minds.

At one point, I glanced forward a couple yards to discover a timber rattlesnake coiled on the path directly in front of us! Our hearts jumped as we immediately froze on the trail, taking in the awesomeness of the snake that had its eyes fixed on us. Calmly, we stepped back a couple steps. I found a thin, long branch, which I used to lift the snake and place it on a large rock beside the trail. Slowly we left the snake as we chatted with excitement about the privilege we had of encountering this beautiful creature.

> *Consideration needs to be given to how we compare the value of humans to the value of nonhuman creatures.*

So how do you decide if a snake is valuable or not? Here is how our decision was made. We were aware such a sighting was somewhat uncommon, so we reveled in the special moment. Because of our strong appreciation for living things, striking out at the snake would have been counter to our respect for life – and it would not have been wise! Wonder triumphed over any desire to destroy.

Subsequently, we learned snakes can be indicators of a healthy environment. The presence of the timber rattler was a reminder that there was good balance in the ecosystem through which we were walking. The presence of a top-level carnivore, such as the rattler, indicates that the herbivores, the plant eaters, are doing well. The presence of herbivores means the plant life is functioning as it should. Food webs are complex, and we as humans are part of that system. Our role within food webs is one way to engage the concept of ecological or creation care ethics.

At its core, ethics gives definition to what is right or wrong conduct. The study of moral values as applied to creation care ethics involves asking the question, "On what basis do we, as Christians, determine what counts or has value?" Consideration needs to be given to how we compare the value of humans to the value of nonhuman creatures. Do these creatures count only because they supply a human need – or do they have value in and of themselves? Our ecological ethic grows from our beliefs and will be shown through our practices.

An important starting point for our framework is to think about what God values. The creation story in Genesis 1 recounts God's actions in a pattern of days – a rhythm of order and time. Not only does God create, God also steps back and views his work. God declares a value about what has been made through a repeated phrase, "It was good." It is significant to note that God validates the importance of creation – both the nonliving and living things – prior to the presence of humans. The goodness of each created thing was pronounced on the merit of what valued to God and how it reflected the image of the triune God. The goodness was not because it had some use or value for humans.

The last creative act on the sixth day was the forming and bringing to life of humankind in God's image. In Genesis 1:31 we read of God's superlative response to all of creation, which now includes people: "God saw everything that he had made, and indeed, it was very good. And there was evening and there was morning, the sixth day."

But God had one more action to fulfill. This is the act of resting, which adds another critical perspective to our ecological ethic. Rest, renewal, and rejuvenation are part of the value system. This seventh day action is so important God declares a blessing on it, too. So rather than concluding the creation story with the creation of man and woman, which could have implied an elevation of humans to the top position in creation, God stays at the center of all the relationships by including rest in the pattern for life.

> *Our ecological ethic grows from our beliefs and will be shown through our practices.*

God loves the world. Each part of God's creative package matters. Even matter – land and water – matter. Plants. Animals. Humans. Rest. It all mattered to God. The basis of our ethic was established. What matters to God should matter to us.

The love God has for all parts of the world is demonstrated in the stories in the Old Testament as God related to people and the Earth through redemptive acts. The culminating display of God's love for the whole world happened through Jesus Christ taking on human flesh (John 1:14). This act of the Son taking on a physical nature – made of the earth – carries meaning for all of creation since Christ's death and resurrection made reconciliation available for all things (Colossians 1:20). But there is additional significance for our creation care ethic that comes through the incarnate Christ – that is his life. Cal DeWitt notes this importance in his book *The Environment & the Christian*:

> Although Jesus' ecological ethic is not prominent in the text of the Gospels, the fact is that Jesus, as a Jew, lived a profound ecological ethic. He demonstrated a practical way of dealing with problems that is relevant to the solution of any practical problem, including those of contemporary ecological politics.[2]

While we do not have direct statements from Christ telling us to care for the Earth, we see his care toward all things demonstrated repeatedly. As Christ followed the Jewish law, he lived out an ecological ethic through several principles contained in the law. These were the principles of do not destroy, do not inflict pain on living things, and keep the Sabbath. Awareness of these ethical positions adds understanding to how Christ conducted his life. Using this ecological ethic as a model for our lives today would bring restoration as we deal with the environmental crisis. Another key for a creation care ethic – modeled in Christ's life – is his teaching found in Matthew 7:12: "In everything do to others as you would have them do to you; for this is the law and the prophets."

> *The ethic of love is revealed most fully in Christ ...*

Consider the breadth and depth of this teaching. Typically we think of the term "others" as meaning other human beings. While this is a critical aspect, I believe "others" also applies to the rest of the created order. There is an inter-connected relationship described in this context that has congruence with how ecology functions.

An act on one part of an ecosystem or community has ramifications on other parts of that system or community. Valuing things for more than just their utilitarian potential to humans is implied in the broader context of this Matthew passage and in the parallel passage in Luke 6:31. Christ's ethic teaches respect for the "other," for the least of these, for the child, for the enemy, and for the poor. This ethic of love is the purpose of the "law and the prophets." It is also revealed most fully in Christ.

Group questions

1. *How do you describe your relationship with the Earth? As that of friend or foe? Or both/neither? Why do you describe your relationship in this way?*

2. *How do you see Old Testament stories and teachings supporting the ecological ethic described in Genesis 1? Which stories raise questions as to the importance of a creation care ethic?*

3. *In what ways do you understand Christ's life modeling an ethic that includes the importance of stewardship of the Earth?*

Finding our way

In the process of developing our creation care ethic, we are challenged by many other teachings and philosophies. It is important that we acknowledge how we are influenced by the individualism and consumerism of Western culture. On one hand, we are grateful for the freedoms we experience in North America. On the other hand, we know how easy it is for our decisions and ethics to be governed by the culture in which we live. The struggle with the culture in which we are

commingled was also an issue in the early church. Paul teaches about this in
2 Corinthians 4:4-6:

> "In their case the god of this world has blinded the minds of the unbeliev-
> ers, to keep them from seeing the light of the gospel of the glory of Christ,
> who is the image of God. For we do not proclaim ourselves; we proclaim
> Jesus Christ as Lord and ourselves as your slaves for Jesus' sake. For it is
> the God who said, 'Let light shine out of darkness,' who has shone in our
> hearts to give the light of the knowledge of the glory of God in the face of
> Jesus Christ."

Christ followers see the light of the gospel. This good news for the whole
world is described as a distinct contrast to the culture of unbelievers. Because
"light" is missing, minds are blinded. Within society there is recognition of needs
– for example the need for ecological health and restoration – but without the
understanding of who God is and the importance of his revealed image, the
message is hopeless. Having the light and life of Christ shine in our hearts brings
joy and hope to the creation care ethic. What God intends for human conduct
has been revealed most fully in Christ. This foundation informs our ways of caring for creation.

*It is important that we acknowledge how we are influenced by the
individualism and consumerism of Western culture.*

Our beliefs are important in informing our ethical positions. Anabaptist belief
statements give expression to the Mennonite understanding of the Scripture and
have been formed in the context of community. The following is a relevant example:

> God formed them from the dust of the earth and gave them a special dignity
> among all the works of creation. Human beings have been made for rela-
> tionship with God, to live in peace with each other, and to take care of the
> rest of creation.[3]

Note the three relational roles and the ethical implications named in this statement. The first is relationship with God. Our actions are framed in God's image.
The relationship with people is grounded in the ethical perspective of peace. A
parallel third relationship is the ethic of faithful caregiving to all of creation. A
potential challenge to our ecological ethic is found in the first sentence of this
quote. We can fall into the trap of interpreting "special dignity" to mean the right
to dominate all the other works of creation. The ethical problem with this view is
that of operating from a position of anthropocentrism – human centeredness.

The earlier commentary on who Christ is – and the ethic that he taught through his life, describes a Christocentric – Christ-centered – position. If we allow humankind to usurp the centrality of Christ, we slide into destructive behaviors in relation to all that God has given us. When we function in an elevated, human-centered role, the image of God is denied. The result is the rest of creation suffers and groans. Heather Ann Ackley Bean writes about the challenges Mennonites have had in articulating a creation care ethic in her essay, "Toward an Anabaptist/Mennonite Environmental Ethic." There are ways in which the Mennonite belief system has moved us toward anthropocentrism, resulting in inconsistent practices. She draws on Scriptures and Anabaptist writers to develop a more coherent understanding. Importantly, she comments on the critical connection between faith and practice.

> The Anabaptist belief that ethics is applied theology is itself a resource for environmental ethics. No abstract ethical theory will suffice in the current ecocidal context. Any environmental ethic must be lived to be effective.[4]

Making choices

A way to understand the ethical challenge of being in balance with nature is to consider a spectrum of views regarding our relationship with the Earth. At one end of the continuum is the concept of a "Wild Earth." This is defined by the ideas of the Earth being untouchably sacred; fewer people and more nature is best; protecting and revering the Earth is of prime importance. At the other end of the spectrum is the understanding of the Earth as "Nature Factory." The main concepts of this view are to see the Earth as a resource for a technological and industrial economy – with the human role to tame, subdue, and extract.

Consider the concept of "Earth as Home" – a place we care for.

I am not comfortable with either extreme. At the same time, I am drawn to each and see how I make choices based on each end of the spectrum. As an alternative ethic, I propose a third alternative. Consider the concept of "Earth as Home" – a place we care for. Defining components include a moral ecology of interacting with nature and practicing concepts like agroecology and sustainability.

Acknowledging our differences across this spectrum is important for any discussion of creation care ethics. Our personal perspectives will reflect multiple aspects described in this simplistic range of views I have given. It is important that we listen to each other as we work to find our common values. As communities of faith, we need to build on the strength that comes out of our core beliefs and values, rather then regress to polarizing views that zap our energy and witness.

Our creation care actions are informed by healthy conversations in community where we explore the foundations of our beliefs. I feel our discussions can be well served by considering the seven points developed by Steve Bouma-Prediger in *For the Beauty of the Earth: A Vision for Creation Care*, which are based on scriptural principles. This list develops a framework and set of values that guide how we are to live out our role as caretakers of the Earth.

1. The various kinds of animals and plants that populate the Earth are created by God and are, therefore, valuable – irrespective of their usefulness to us.

2. The Earth and its creatures are finite.

3. We are limited and often self-deceived in how we view the world.

4. The God-designed world is fruitful and able to sustain itself.

5. Work is good, but so is rest.

6. The Earth is not ours.

7. The cries for righteousness and justice must not go unheeded.[5]

An ecological value that has been part of an Anabaptist heritage is that of simple living. This way of life is based on our dependence on God, not on our commitment and history of frugality. The simple living value is tested in our cultural intersection with individualism and materialism. We have "bought into" consumerism as represented in much of Western society. Many injustices emerge in the local and global contexts because of our lifestyle choices. Heather Ann Ackley Bean calls us to confession and a change in behavior:

> Truly sorrowful contrition requires that we know and believe in the reality of ecological crisis and that we fully accept our share of responsibility for it. Knowledge of sin and admission of guilt are still one step removed from true repentance – actual change of behavior. A practical conversion, from environmentally destructive behavior to ecologically sustainable behavior, must be the goal of … environmental ethics.[6]

I believe we can be an effective witness to the good news of Christ by seeking out and practicing stewardship of God's creation. The results are an honoring of the Creator and a profound and attractive influence on the society in which we live.

Prayer:

Oh God, our Creator, Redeemer, and Sustainer,

We give you thanks for forming us and all of creation in your image.

Thank you for revealing yourself to us through the creation around us

and by sending Christ to Earth in human flesh.

We struggle with finding and living an ecological ethic that is truly life-giving.

We admit we have failed in being faithful caregivers of your Earth.

We have centered on ourselves rather than giving first place to you.

We rejoice in your redemption.

You have graciously given us life and given us life again.

Fill us with your wisdom as we choose the path of actions and behaviors that

are ecologically sustainable.

Amen.

End questions

1. *How do you see yourself working within your community of faith to better understand and develop a creation care ethic? What are some of the desired outcomes of this work?*

2. *How does Scripture support the seven points for an ecological ethic outlined by Steven Bouma-Prediger?*

3. *Reflect on Heather Ann Ackley Bean's statement of a need for a "practical conversion, from environmentally destructive behavior to ecologically sustainable behavior." What is the relevance for you in your daily life?*

An activity to do at home:

> *Look though your checkbook – a record of what is valuable to you – and make a list of how your ecological ethic is reflected by your expenditures.*

Sources

1   James Jones, *Jesus and the Earth* (London: Society for Promoting Christian Knowledge, 2003), 64.

2   Calvin B. DeWitt, *The Environment & the Christian: What Can We Learn from the New Testament?* (Grand Rapids: Baker Book House, 1991), 96.

3   "Article 6: Creation and the Calling of Human Beings," in *Confession of Faith in a Mennonite Perspective* (Scottdale, Pa.: Herald Press, 1995), 28.

4   Heather Ann Ackley Bean, "Toward an Anabaptist/Mennonite Environmental Ethic," in *Creation and the Environment: An Anabaptist Perspective on a Sustainable World*, ed. Calvin Redekop (Baltimore: Johns Hopkins University Press, 2000), 198.

5   Steven Bouma-Prediger, *For the Beauty of the Earth: A Christian Vision for Creation Care* (Grand Rapids: Baker Academic, 2001), 159.

6   Heather Ann Ackley Bean, "Toward an Anabaptist/Mennonite Environmental Ethic," in *Creation and the Environment: An Anabaptist Perspective on a Sustainable World*, ed. Calvin Redekop (Baltimore: Johns Hopkins University Press, 2000), 204-205.

Ecojustice

"Ecojustice is concerned with reversing the ecological crisis and at the same time securing justice for the poor. It's about responsible earth stewardship as well as responsible person stewardship." — Art and Jocele Meyer[1]

The voice of the waters

Good water from a well is so refreshing! But what do you do when water is polluted? And whose problem is it anyway? A group of young adults raised these questions while interacting with a geologist as they viewed a groundwater model. The geologist's model consisted of a tall thin box with two glass sides. Between the glass walls were layers of soil, sand, and gravel, which represented a cross section of the Earth. A tube had been inserted through all the layers, which served as a well, reaching into the aquifer of the model. A watering can was used to provide "rain" on the soil at the top of the model. By looking through the glass sides, the students could see how the rainwater recharged the aquifer even while water was "pumped" to the surface of the model. This was a healthy water system.

Slyly, the geologist added three drops of red food coloring onto the soil at the top of the model. As rain continued to fall, the dye slowly made its way to the aquifer. Once the food coloring was in the water table, the water being drawn from the well turned red. Added rainfall diluted the color, but the water continued to have a red tint.

The students and the geologist discussed how this happens repeatedly across the landscape as toxins and pollutants are dumped into our water systems. As they talked, the students could see the pinkish water in a pan that had been collected from the model's well — and the red color in the sand and gravel. At the end of the demonstration, several students gasped as the geologist casually (and intentionally) took the pan of water and threw the contents out on the grass. This ironic act of the geologist emphasized how knowing what is right doesn't always turn into right actions.

... knowing what is right doesn't always turn into right actions.

As an observer, I was vividly reminded of the challenges of environmental degradation and how actions in one part of the landscape have broad ranging impact in other locations. In my mind I could see the factories, homes, and businesses releasing various kinds of pollutants that would spread through aquifers, the atmosphere, body tissues, rivers, and soil. Withdrawing the pollutant once it is released is very difficult, if not impossible. These actions result in injustices against people and all of creation.

The study of ecojustice in this chapter is a connecting link between the previous chapters on theology and ethics of creation care and the following chapters, which deal with specific issues of injustice and ecological challenge. As we gain insight into our beliefs and values about caring for the Earth, our alertness to the inequities within the Earth's systems increases. This also calls us to action, otherwise our beliefs and values have little meaning.

Ecojustice emerges out of the intersection of the concepts of social justice and ecological sustainability. As Christians, we believe our faith in Jesus Christ is followed by actions that represent what he would do. In Matthew 25:35-36, Jesus describes his experience with faithful followers who had acted in such instinctive ways that they were uncertain of when the events had occurred.

> "For I was hungry and you gave me food, I was thirsty and you gave me something to drink, I was a stranger and you welcomed me, I was naked and you gave me clothing, I was sick and you took care of me, I was in prison and you visited me."

In obedience to this teaching – and many others in the New Testament – I believe taking social justice issues seriously is part of who we are. We are also aware that caring for all of God's creation is in keeping with loving what God loves. What matters to God ought to matter to us as Christians. Ecological sustainability is part of our larger responsibility. Here we embrace the need to seek patterns that bring life and health to the Earth's systems both for the present and for the generations to come.

Ecological sustainability is part of our larger responsibility.

Justice and sustainability involve practices that are circular and interlinked. One cannot be had without the other. Justice for people promotes sustainability in all of creation. Sustainability makes justice possible. It is important to think about the linkages and cycles that hold these two concepts together. Ecojustice represents both concepts. Consider the similar cyclical nature of faith and works described in James 2:14-17:

> "What good is it, my brothers and sisters, if you say you have faith but do not have works? Can faith save you? If a brother or sister is naked and lacks daily food, and one of you says to them, 'Go in peace; keep warm and eat your fill,' and yet you do not supply their bodily needs, what is the good of that? So faith by itself, if it has no works, is dead."

The combined importance of faith and works in our Christian life is well described in James. One of our faith/works is that of ecojustice – the combination of ecological sustainability and social justice. Given our earlier study on creation care ethics, I believe we could include "tilling and keeping" actions toward the Earth right alongside the caring for our brothers and sisters within this passage from James. As Steven Bouma-Prediger writes,

> Our passion for justice should embrace all creatures – for their sake and for the sake of their human neighbors whose voices cry out for justice to roll down like waters and righteousness like an ever flowing stream.[2]

Our responsibilities include acting justly in human communities and restoring ecosystems locally, regionally, and globally. City, county, state, and national borders do not stop the injustices caused by the consumptive practices of human beings. Hazardous wastes leak into the aquifers under towns and cities. These toxic materials move throughout an aquifer, polluting the wells in the region.

Our responsibilities include acting justly in human communities and restoring ecosystems locally, regionally, and globally.

Poisonous gases are released into the air from factories. People in surrounding communities breathe the polluted air and experience an increased incidence of cancer. Methane escapes from our landfills. This greenhouse gas rises into the atmosphere, contributing to climate change across the globe. Soil erosion, fertilizers, and pesticides from agricultural practices contaminate lake and river systems extending far from the source. The burning of coal to produce electricity in one region sets up the conditions for acid rain to fall in another region. People and organisms in distant locations suffer the consequences from all these actions and many more.

We are all affected. But it is the poor, the marginalized, and the powerless who take the brunt of the injustices, both social and ecological. The reality is that our North American lifestyle generates inequity and violence both nationally and internationally. Environmental degradation and injustices to people are linked. The cries for mercy and justice are loud, but often unheard or ignored. Brian Walsh and Sylvia Keesmaat point out the unheard voices in their study of Colossians.

> The one voice that is always drowned out in our cultural cacophony – until it screams at us through ecological disasters – is the voice of creation. It takes a certain meekness and receptivity to hear that voice. A community renewed in knowledge according to the image of the Creator embraces meekness as integral to its way of engaging the world, because it follows a Lord who said, "Blessed are the meek," and who was recognized as the Messiah by the very stones on the side of the road (Matthew 5:5, Luke 19:40).[3]

I believe that, as Christians, we are called to hear the voices crying out for justice. We should be the voice for those who have no voice. Our actions are to reflect the image of our Creator.

Group questions

1. *Respond to the concept that there is a cyclical link between social justice and ecological sustainability. In what ways do you agree and/or disagree with this idea?*

2. *Read James 2:8-26. How do the faith and works principles in this passage relate to ecojustice values of creation care?*

3. *What are some voices within the ecojustice realm that are not being heard?*

Three prophetic voices

Climate change is a reality that is affecting all people – especially the poor and vulnerable of the world. The Intergovernmental Panel on Climate Change states in its most recent report,

> Warming of the climate system is unequivocal, as is now evident from observations of increases in global average air and ocean temperatures, widespread melting of snow and ice, and rising global average sea level.[4]

The reality of climate change has been carefully and clearly documented. Multiple organizations around the world are taking note of the ways in which the most vulnerable are affected by these climatic changes. For example, the United Nations Human Development Reports acknowledges that,

> Climate change is the defining human development challenge of the 21st Century. Failure to respond to that challenge will stall and then reverse international efforts to reduce poverty. The poorest countries and most vulnerable citizens will suffer the earliest and most damaging setbacks, even though they have contributed least to the problem.[5]

As a people seeking God's righteousness, we are called to respond to the cries of those without a voice. God's desire is to see a reconciliation of all things, and we are to be participants in this plan. Making things right is a central part of this plan. N. T. Wright articulates this well in his book *Simply Christian: Why Christianity Makes Sense*.

> God does indeed intend to put the world to rights. There is a cry for justice, which wells up from our hearts, not only when we are wronged but when we see others being wronged. It is a response to the longing, and the demand of the living God that his world should be a place not of moral anarchy, where the bullies always win in the end, but of fair and straight dealings, of honesty, truthfulness, and uprightness.[6]

This sets the stage for us to listen to three prophetic voices, each with a call to action. Ecojustice is to be pursued.

> *People seeking God's righteousness are called to respond to the cries of those without a voice.*

The Old Testament prophets

The prophets of the Old Testament were keenly aware of the inequity and injustice that had pervaded the lives of the people of God. They did not hold back in their messages as they pleaded for believers to change their ways. In Jeremiah 7:5-7, the weeping prophet implores the people to pay attention to how they have been treating others and to choose a God-honoring response.

> "For if you truly amend your ways and your doings, if you truly act justly one with another, if you do not oppress the alien, the orphan, and the widow, or shed innocent blood in this place, and if you do not go after other gods to your own hurt, then I will dwell with you in this place, in the land that I gave of old to your ancestors forever and ever."

God's message is that we are to assist in putting the world to rights. This is played out in our treatment of one another, of our neighbors near and far, and of the most vulnerable. I believe this also includes the dimension of ecological rightness or sustainability. Jeremiah's declaration draws attention to health in the land when justice happens for all people. This is the cyclical interconnectedness of ecojustice.

> *A journey through the prophets of the Old Testament is a reminder of how fundamental justice, righteousness, and mercy are to the heart of God.*

A journey through the prophets of the Old Testament is a reminder of how fundamental justice, righteousness, and mercy are to the heart of God. Over and over the prophets ask for a new heart and spirit (Ezekiel 11:19), a knowing of God's ways (Hosea 6:3), a seeking of good and not evil (Amos 5:14), and a humble walk with God (Micah 6:8). Heeding the voices of the prophets forms the seedbed for ecojustice, as noted in the poetic words of Amos 5:24: "But let justice roll down like waters, and righteousness like an ever-flowing stream."

The good news of Jesus

Jesus began his three years of earthly ministry with a dramatic reading from Isaiah. Listen to the message from Luke 4:18-19 that Jesus shared in the synagogue.

> "The Spirit of the Lord is upon me,
> because he has anointed me
> to bring good news to the poor.
> He has sent me to proclaim release to the captives
> and recovery of sight to the blind,
> to let the oppressed go free,
> to proclaim the year of the Lord's favor."

Jesus emphasized that the good news was for the most vulnerable, the power-less, the rejected, and the oppressed. This is a message of hope for all. But hope for all can also be disturbing. After Jesus made comments on the text – comments that challenged the status quo of the listeners – the audience became angered and drove him out of town. Hope and justice for all can be uncomfort-able for those of us who already have much. We are to let go of our material security in order for the poor and oppressed to have what they need and to be set free.

Jesus continued to bring this tension-inducing good news message to light throughout his ministry. People were attracted to him because of his compassion, hope, and healing. But listeners also chaffed under his direct statements toward those with power and wealth who were dismissive of those who did not have these resources. Jesus' life and teachings show us time and again what is needed to "put the world to rights."

> *The peace God intends for humanity and creation was revealed most fully in Jesus Christ.*

Earlier we reflected on the Matthew 25 story from the end of Jesus' three-year ministry. The concluding remarks to the faithful servants in Matthew 25:40 offer a key understanding for social justice and ecological sustainability: "And the king will answer them, 'Truly I tell you, just as you did it to one of the least of these who are members of my family, you did it to me'."

In practicing ecojustice, we must pay attention to the least of these. All members of God's family – the whole of creation – are included. The faithful servants knew the value of all members rather than focusing on comfort for self.

Anabaptist peace

The final prophetic voice we will note in our study of ecojustice is one that has emerged out of scriptural discernment by Anabaptists. I am grateful for the succinct statement on ecojustice found in the *Confession of Faith in a Mennonite Perspective*: "The peace God intends for humanity and creation was revealed most fully in Jesus Christ."[7]

In applying our understandings of peace and justice, we typically focus on what that means in relation to human beings – our neighbors next door and around the world. While this is important, we can easily fail to notice that God's intended peace is also for creation. Within this single statement we have the merger of social justice and ecological sustainability. How poignant! Ecojustice is part of our call to be faithful Christians. I am moved to understand and apply this concept especially because it is "revealed most fully in Jesus Christ."

As a community of believers, we are to seek peace and pursue it. This includes addressing the tragedy of war and ecological devastation, the lack of water for people and loss of biodiversity, conflict with neighbors and wanton pollution, and overlooking the poor and destroying habitat.

> *Our ministries as a church should include mediation, reconciliation, and nonviolent resolution within the realm of ecojustice.*

God enters our world through Christ – the Creator, Redeemer, and Sustainer. We are to participate in Christ's ministry of peace and justice. He has called us to find our blessing in making peace and seeking justice.[8] In response to Christ's way of peace, our ministries as a church should include mediation, reconciliation, and nonviolent resolution within the realm of ecojustice. Our actions toward all of creation should bless, redeem, and restore. As ministers of ecojustice, we need to break with the familiar Western norms of materialism and greed. Through Christ we offer an invitation to wholeness.

Prayer:

Lord of the Universe, Lord of our lives,

We see your truth and justice displayed through love and kindness.

We thank you for the clear voice of the prophets, Jesus, and our heritage.

But the messages are uncomfortable.

How should we live? How can we be faithful servants who bring peace?

We easily turn a deaf ear to those with weak voices.

We know there are many situations where there is no utterance.

We ask for your wisdom, exhortation, and strength to live the life of peace.

May our life of peace include acting with justice toward all of creation.

Teach us to love as you love.

Amen.

End questions

1. Read through the noted texts from the Old Testament prophets. What are the messages of ecojustice you find?

2. What do you observe in the life of Christ that speaks to both the need for social justice and ecological sustainability?

3. What are the important applications that are possible when we recognize God's peace is for both humanity and creation?

An activity to do at home:

Read a news report or watch a news program that describes a conflict between people. It could be local, regional, or international. What social justice issues do you observe in the story? What are the ecological components that may be noted?

Sources

1   Art and Jocele Meyer, *Earthkeepers: Environmental Perspectives on Hunger, Poverty, and Injustice* (Scottdale, Pa.: Herald Press, 1991), 33.

2   Steven Bouma-Prediger, *For the Beauty of the Earth: A Christian Vision for Creation Care* (Grand Rapids: Baker Academic, 2001), 169.

3   Brian Walsh and Sylvia Keesmaat, *Colossians Remixed: Subverting the Empire* (Downers Grove: InterVarsity Press, 2004), 194-195.

4   Intergovernmental Panel on Climate Change, "Summary for Policymakers of the Synthesis Report of the IPCC Fourth Assessment Report, Draft Copy 16 November 2007," http://www.ipcc.ch, 2007.

5   United Nations Human Development Reports, "Fighting climate change: Human solidarity in a divided world," http://hdr.undp.org/en/reports/global/hdr2007-2008, 2007.

6   N. T. Wright, *Simply Christian: Why Christianity Makes Sense* (New York: HarperCollins, 2006), 225.

7   "Article 22: Peace, Justice, and Nonresistance," in *Confession of Faith in a Mennonite Perspective* (Scottdale, Pa.: Herald Press, 1995), 81.

8   "Article 22: Peace, Justice, and Nonresistance," in *Confession of Faith in a Mennonite Perspective* (Scottdale, Pa.: Herald Press, 1995), 81.

Creation Care and Economics

"For our environmental future to be secured, the economic pendulum must swing back from the impersonal, individualistic global economy toward an era of renewed cooperation within strengthened local communities. Insofar as humans desire environmental justice, we can still make a difference."
— *James Harder and Karen Klassen Harder[1]*

A bee economy

I always felt a twinge of excitement whenever I put on my beekeeping hat, veil, and gloves. I was heading out to explore a world of wonder and hidden spaces. What a delight to peer into the hive and see such astonishing activity. Stings were few during the 10 years I had several hives of bees while living in western Pennsylvania. And the honey was sweet!

The complexity of the hive was fascinating to me, both in terms of the structure of the wax comb and the work of the community. Imagine a "bee accountant" collecting information on the colony to make sure the economic business plan was working. Here is some of the information he would have gathered. The hive needed to have at least 30,000 workers to be productive, but only one queen. The workers will fly one to two miles away from the hive to forage on flowers, even going eight miles if necessary. One worker bee will make about one-twelfth of a teaspoon of honey in her lifetime. Foragers must collect nectar from about two million flowers to make one pound of honey. Bees will consume 17-20 pounds of honey to produce one pound of beeswax. And to think that I would take 40 pounds of honey from one hive each year!

The "bee accountant" for the hive isn't counting economic values that are important to agriculture. Pollination by bees improves crop yield and quality by approximately 14 billion dollars annually. In one small way, this represents the intersection between nature's systems and our human systems of economics.

The word "economy" comes from the Greek word *oikonomos*, which combines *oikos* meaning house or household and *nemo* meaning manage. This awareness sets up an important realization for our study of creation care and economics. As we have observed, ecology and economics have the same root prefix – house or household. The one is the study of the house and the other is the management of the house. We also know that stewardship at its rudimentary level means "keeper of the house." The interconnected connotations of these words provide an excellent framework for healthy creation care stewardship. On the other hand, there is a loss of vitality when the word concepts function in ways that are at odds with each other.

Our concern for the systems of nature has gone

Wendell Berry has spent much of his lifetime thinking and writing about this tension. He is quick to note how our Western civilization has failed to keep economics and ecology in a proper equity.

> The "environmental crisis" has happened because the human household or economy is in conflict at almost every point with the household of nature. We have built our household on the assumption that the natural household is simple and can be simply used. We have assumed increasingly over the last five hundred years that nature is merely a supply of "raw materials," and that we may safely possess those materials merely by taking them.[2]

Our concern for the systems of nature has gone awry. The treatment of nature's household as a cupboard to be consumed until the cupboard is bare, is not a sustainable model. Nor does it represent a creation care ethic. The basis of our current economy is consumer-driven. Without giving rightful place to ecological health, this translates into a consumptive lifestyle and an environmental crisis.

While it is typical for us to use the term "consumer" to mean "one who buys," the root word "consume" carries the defining concepts of using up, eating, wasting, and destroying. This may seem harsh to focus on understanding "consumer," but it is true to our North American lifestyle. The following information from the Worldwatch Institute's *State of the World* report brings our understanding of "consumer" into focus.

The United States and Canada share 31.5 percent of the world's private consumption expenditures while only having 5.2 percent of the world's population. In great contrast, Sub-Saharan Africa's private consumption expenditures are 1.2 percent of the world's total while having a 10.9 percent share of the world's population.[3]

In the United States today, there are more private vehicles on the road than people licensed to drive them. The average size of refrigerators in U.S. households increased by 10 percent between 1972 and 2001, and the number per home rose as well. New houses in the United States were 38 percent bigger in 2000 than in 1975, despite having fewer people in each household on average. Because of these consumption patterns, the United States, with just 4.5 percent of the world's population, releases 25 percent of global carbon dioxide emissions. Yet increased consumption has not brought Americans happiness. About a third of Americans report being "very happy," the same share as in 1957, when Americans were only half as wealthy.[4]

We have a consumptive problem.

We have a consumptive problem. A challenge for us in the North American church is the lifestyle that is represented behind these statistics. It is one with which we are comfortable. In fact, it is the norm in our homes and in our churches. This is a reality we do not like to face. Consider 1 Timothy 6:6-10 where Paul sets forth the standard called contentment.

> "Of course, there is great gain in godliness combined with contentment; for we brought nothing into the world, so that we can take nothing out of it; but if we have food and clothing, we will be content with these. But those who want to be rich fall into temptation and are trapped by many senseless and harmful desires that plunge people into ruin and destruction. For the love of money is a root of all kinds of evil, and in their eagerness to be rich some have wandered away from the faith and pierced themselves with many pains."

This is a sobering passage for us today. We know we represent the rich in the world. While it is not money that is evil, the reality is our North American wealth has given us great comfort – at a high cost to our neighbors around the globe and the rest of creation. The prophets call out repeatedly about this inequity and

the destructive management of the household. In Isaiah 24:4-5, the prophet
speaks to the desolation that comes not only to people, but also to other parts
of creation.

> "The earth dries up and withers,
> the world languishes and withers;
> the heavens languish together with the earth.
> The earth lies polluted
> under its inhabitants;
> for they have transgressed laws,
> violated the statutes,
> broken the everlasting covenant."

*Our North American wealth has given us great comfort – at a high
cost to our neighbors around the globe and the rest of creation.*

Our task is to seek and then practice a way of life that truly represents our
respect for the natural order of creation and justice for all people. We delight in
God's care and keeping when we are experiencing plenty. It is difficult when the
reciprocal occurs. Can we be content if our comforts are taken away, or if we
choose a lifestyle that is not driven by consumption? Again, Wendell Berry sounds
the alarm about our current dilemma.

> And so we have before us the spectacle of unprecedented "prosperity" and
> "economic growth" in a land of degraded farms, forests, ecosystems, and
> watersheds, polluted air, failing families, and perishing communities.[5]

Group questions

1. *In what ways do you see North American consumption and materialism directly tied to the environmental issues we face today?*

2. *What are the teachings from Isaiah 24 and 1 Timothy 6 that may be included in a well-managed household/economy?*

3. *How might learning "to respect the natural order of creation" help us as Christians to be more committed to contentment rather than consumer-driven comfort?*

A changed economy

The honeybees I kept were generous. Every year I was able to remove honey for our family's use. But I knew there was a limit in how much honey I could remove. If I took too much, the colony would not survive the winter. Knowing how much was enough is a principle that should be applied to our understanding of a healthy economy. The results of a consumptive lifestyle – one that takes too much – looks like a beehive that did not survive the winter. The way to healing for the brokenness caused by our North American lifestyle is outlined in Isaiah 58:6.

> "Is not this the fast that I choose:
> to loose the bonds of injustice,
> to undo the thongs of the yoke,
> to let the oppressed go free,
> and to break every yoke?"

Restorative actions are found in the verses that follow. Give bread to the hungry. Provide a home for the homeless. Clothe the naked. Reach out to our kin – all of humanity. Two causes of the brokenness are recorded in verse 13. One is "trampling the Sabbath." This represents disrespect for God's created order. Remember that God concluded the creation events with rest, which kept God in the center rather than humans. The second cause is "pursuing your own interests." Our self-interests revolve around wants. Here wants become defined as needs and form a basis for consumption. Meeting personal interest becomes the prime motivation at the cost of the whole planet.

> *Knowing how much was enough is a principle that should be applied to our understanding of a healthy economy.*

Living in North America in the 21st century is easy – and it is hard. Following the masses in a materialistic, consumptive way of living is the norm. Continuing in this direction is comfortable because it is what we know best – and it is hard to buck this flow. Even when this direction is built on debt – the debt of credit cards, the national debt, the debt of inequity, and the debt of a degraded environment. David Orr underscores this quandary in his chapter titled, "The Ecology of Giving and Consuming," in *The Nature of Design*.

> Excess has become the defining characteristic of the modern economy, evidence of design failures that cause us to use too much fossil energy, too many materials, and make more stuff than we can use well in a hundred years. If, however, we intend to build durable and sustainable communities, and if we begin with the knowledge that the world is ecologically complex, that nature does in fact have limits, that our health and that of the natural world are indissolubly linked, that we need coherent communities, and that humans are capable of transcending their self-centeredness, a different design strategy emerges.[6]

Orr describes the importance of rising above and going beyond our centeredness on self. This is the same prophetic statement we noted in Isaiah 58:13. It is only when we come to this juncture that we are able to design and choose a different strategy for living. With this repentant action we can "be called the repairer of the breach, the restorer of the streets." (Isaiah 58:12) James Harder and Karen Klassen Harder present the urgency of this concept in their essay, "Economics, Development, and Creation."

> An emerging alternative economic vision suggests that the drive for unrestrained economic growth itself has become the most important problem facing humanity. The primary concern is that a world of over six billion people striving for material satisfaction is drawing ever more heavily from finite supplies of natural resources to fuel an economic growth model destined to lead to an ecological disaster and global poverty without precedent. There is an urgent need to find an economic way of life that is environmentally sustainable.[7]

The Harders identify three myths followed in our current economy. Understanding these to be myths aids in our thinking about the kinds of changes that are required to move against the economic status quo.

Myth 1: Growth equals development.

It is often assumed that growth means positive change, but it can just as well cause a system to be out of balance. Development occurs when the goal is to achieve a desired quality that is beneficial to the health of the whole.

Myth 2: There is no limit to growth.

While we understand the resources of the Earth are limited, we practice a way of life that hopes there will always be more. Ever-expanding economic growth is not possible since the Earth has limits to its carrying capacity for life. We are at a point in time when we are beginning to acknowledge that there are too many demands on the biosphere.

Myth 3: Growth resolves poverty.

The production of human-defined wealth is at an all-time high, but there is little evidence that the decades of this artificial economic growth has trickled down to the poor. As the issues of climate change loom on our horizon, we are more aware of how outcomes of consumptive patterns have the greatest impact on the poor. Without an understanding of limits, we are faced with an increased poverty for humankind and for all of creation.

> *While we understand the resources of the Earth are limited, we practice a way of life that hopes there will always be more.*

Shaking these myths from our approach to living requires a radical rethinking of how we live in North America. As a church, Mennonites have taken this issue seriously through the work we do with Mennonite Central Committee, MMA Praxis Mutual Funds, Mennonite Economic Development, and our mission agencies. We are making right choices and implementing good programs, but they are often at a distance from our day-to-day living. The Harders define a critical shift in our practice.

> The new economy must distinguish needs from wants and require simple living in which consumption is limited. It must thrive on a feeling of sufficiency rather than unlimited desires. It must view nature through the lens of stewardship rather than ownership.[8]

The Earth does belong to the Lord (Psalm 24:1). By shifting our thinking away from a tight-fisted ownership to open hands serving all of creation, we are able to see the benefits of a different way of thinking. Simplicity in living is a key, which we will continue to develop in future chapters. A way of living that has been the norm in other generations is that of giving increased value to local community economics.

Functioning with contentment in a more localized community is a form of self-limitation. We have observed that without valuing limits we become increasingly consumptive. There are many benefits from an economy within a local community. We know where the products come from that we use. We know the people who produced the goods. There is a greater awareness of toxins that may have been used in making the items because the toxins have localized impact. We will understand the human and ecological costs of producing the products, as well as the waste disposal process. These direct connections play a major role to having a healthy and holistic place for living. There will likely be less "stuff," and with less "stuff," simple, contented living becomes the norm.

Functioning with contentment in a more localized community is a form of self-limitation.

We will continue to live in a national and global context even as we place more value on the local/regional economy. Sylvia Keesmaat and Brian Walsh advance principles of how Christian communities can be an influence within the larger systems. They write about this in terms of empire — systems that are powerful and show little regard for the poor and vulnerable. I appreciate how they vigorously name the importance of a shift in our thinking and our behaviors.

In a culture of ubiquitous graven images and rampant consumerist idolatry, we need Christian practices in business, environmental protection, and politics that will topple the idols and energize an alternative economics of God's kingdom.[9]

Jesus, too, spoke with radical ideas about the shift that is needed in our faith practice. The Sermon on the Mount is filled with statements that shook the empire then – and still shakes empires today. His economy was designed around creation's dependence on the Creator. After stating the relationship God has with birds and flowers, he concludes with a summary question that is a critical perspective for the Christian in the 21st century. "But if God so clothes the grass of the field, which is alive today and tomorrow is thrown into the oven, will he not much more clothe you – you of little faith?" (Matthew 6:30)

Prayer:

O God, maker of the household,

Your management of our home – the Earth – is beyond comprehension.

You know all the patterns, needs, rhythms, and interconnections personally.

They all reflect your image.

As ones made in your image, we seek to understand how to be

the best managers of what you have entrusted to us.

Forgive us for grasping space on the Earth with greedy ownership.

Teach us the skills of being stewards.

Grant us your wisdom and strength to move

against the current of consumerism.

May we be people who bring wholeness and restoration

in keeping with your image.

Amen.

End questions

1. *In what ways do you see overcoming the three myths named by the Harders as important in changing our consumptive economy?*

2. *How might we be restorative and regenerative in our local community?*

3. *Discuss the ways that Jesus' teaching in Matthew 6:25-33 is vital for shifting our economic views and values today.*

An activity to do at home:

Make a list of the principles that guide you in the way you care for your home. (Notice that I did not say house.) Then ask:

1. *What does this list indicate about why you view caring for your home as important?*

2. *Which of these principles do you think would be good values for an economy to follow?*

Sources

1   James Harder and Karen Klassen Harder, "Economics, Development, and Creation," in *Creation and the Environment: An Anabaptist Perspective on a Sustainable World*, ed. Calvin Redekop (Baltimore: Johns Hopkins University Press, 2000), 26.

2   Wendell Berry, "The Total Economy," in *Citizenship Papers* (Washington, D.C.: Shoemaker & Hoard, 2003), 63.

3   "State of the World 2004: Consumption by the Numbers," http://www.worldwatch.org/node/1783. The Worldwatch Institute is an independent research organization that works for an environmentally sustainable and socially just society, in which the needs of all people are met without threatening the health of the natural environment or the well-being of future generations.

4   "State of the World 2004: Richer, Fatter, and Not Much Happier," http://www.worldwatch.org/node/1785.

5   Wendell Berry, "The Total Economy," in *Citizenship Papers* (Washington, D.C.: Shoemaker & Hoard, 2003), 66.

6   David W. Orr, "The Ecology of Giving and Consuming" in *The Nature of Design: Ecology, Culture, and Human Intention* (New York: Oxford University Press, 2002), 179-180.

7   James Harder and Karen Klassen Harder, "Economics, Development, and Creation," in *Creation and the Environment: An Anabaptist Perspective on a Sustainable World*, ed. Calvin Redekop (Baltimore: Johns Hopkins University Press, 2000), 3.

8   James Harder and Karen Klassen Harder, "Economics, Development, and Creation," in *Creation and the Environment: An Anabaptist Perspective on a Sustainable World*, ed. Calvin Redekop (Baltimore: Johns Hopkins University Press, 2000), 24-25.

9   Brian Walsh and Sylvia Keesmaat, *Colossians Remixed: Subverting the Empire* (Downers Grove: InterVarsity Press, 2004), 89.

Ecological Footprints

7

"Each person can make a difference because one small positive act multiplied millions of times produces immense benefits." — *Patrick Gonzalez, Climate Change Scientist, The Nature Conservancy, 2007.*

Ecological assessment

Our earlier studies have engaged our hearts and minds in the concepts of theology and ethics related to creation care. From the basis of the ethical understandings, we examined the meaning of ecojustice and ecological economics. In each topic, there was an awareness of putting our beliefs and values into practice. With this chapter we begin a series of lessons that form a framework for examining our current ecological impact and for working to discern how change can best occur in our daily lives.

An important first step is to assess our current lifestyle in an effort to know what kind of effect we are having on the environment. Every organism needs a certain amount of energy to sustain life. All organisms also produce waste. The combination of resource requirements and waste disposal means each organism leaves a mark within its setting. This is neither good nor bad, but it is reality. Ecologists study these relationships, recognizing cycles of life and death, increase and decrease, use and reuse. They look for healthy balance between organisms. The interconnectedness is a critical factor, as no creature functions in isolation from other organisms, resource supplies, and waste streams.

> *An important first step is to assess our current lifestyle in an effort to know what kind of effect we are having on the environment.*

Humans function in similar ways within this ecological mix. We have been blessed with minds to think about these relationships. Our free will allows us to make choices that may benefit or harm other people, as well as ecological systems. God moves in our spirits in ways that convict us to act with wisdom and justice as keepers of the Earth. The special qualities with which we have been endowed are acclaimed by the Psalmist in Psalm 139:14.

> "I praise you, for I am fearfully and wonderfully made.
> Wonderful are your works; that I know very well."

These eloquent words are within an ecological context. David's vantage point for knowing his innermost being was within the rhythms of nature and the vast knowingness of God. The wonderful works of this psalm are not just the marvels of the human body, but the awesomeness of all creation. God knew David well, but David also knew the works of God well. The interactive dynamic of these two forms of knowing are an excellent model for us. Many Christians in North America are detached from knowing the wonder and awe of God's creation. Growing in our understanding of the wonderful works of God is an important starting point in our ecological assessment.

One way to improve our awareness of how we impact Earth's systems is through calculating our ecological footprint. Imagine yourself walking along a sandy beach or through the woods after a snowfall. Each step you take leaves an impression. Counting the number of footsteps that you took would sum up your journey. The evidence is clearly marked out by your tracks in the sand or snow. In an expanded way, our lives leave a set of footprints. Part of the footprint comes from the resources it takes to sustain our lifestyle. Another part is what remains after using a resource. This footprint is more complex to calculate as we try to account for all the inputs and outputs from our living.

In an expanded way, our lives leave a set of footprints.

A challenge to our thinking is that we do not know where many items we depend on come from, nor do we know where the resulting waste goes. Our limited awareness is often similar to the children who when asked, "Where does milk come from?" respond by saying, "From the store." The net is cast far and wide to gather what we use in North America. Once we start counting our footprints – food, water, shelter, transportation, energy sources, etc. – we can be daunted by the complexity of the system that sustains us.

Fortunately, calculators have been developed that assist us in determining our ecological impact. Researchers have collected information on various resource uses and quantities consumed or disposed of in countries around the world. This data has been evaluated in order to know the level of impact that occurs when individuals make lifestyle choices. The researchers have selected a brief list of questions that when answered become a good indicator of a person's footprint on the Earth. Many online calculators may be found by doing an Internet search for "ecological footprint." An example is the Ecological Footprint Quiz found on the Earth Day Network Web site (http://www.earthday.net).

The following quiz questions represent four footprint categories: food, goods, shelter, and mobility.

1. How often do you eat animal-based products (beef, pork, chicken, fish, eggs, dairy products)?

2. How much of the food you eat is processed, packaged, and not locally grown (from more than 200 miles away)?

3. Compared to people in your neighborhood, how much waste do you generate?

4. How many people live in your household?

5. What is the size of your home?

7. Do you have electricity in your home?

8. On average, how far do you travel on public transportation each week (bus, train, subway, or ferry)?

9. On average, how far do you go by motorbike each week (as a driver or passenger)?

10. On average, how far do you go by car each week (as a driver or passenger)?

11. Do you bicycle, walk, or use animal power to get around?

12. Approximately how many hours do you spend flying each year?

13. How many miles per gallon does your car get? (If you do not own a car, estimate the average fuel efficiency of the cars you ride in.)

14. How often do you drive in a car with someone else, rather than alone?

The totals are often given in the number of biologically productive acres it would take to sustain the lifestyle choices. Another unit of measure is the number of earths it would take to meet all the needs if everyone on the Earth lived at the same level as you. According to the Ecological Footprint Quiz, the average person in the United States requires 24 acres, while the worldwide average is only 4.5 acres per person.[1] Jason Venetoulis and John Talberth in their report titled "Ecological Footprint of Nations: 2005 Update," write that the data indicates the consumption of the world's population has exceeded the amount of resources on the Earth.

> For the first time, we have found that footprints associated with crop land, built space, marine and inland fisheries are not sustainable. We found that on a global level, humanity is exceeding its ecological limits by 39 percent – nearly double the amount of ecological overshoot found in our 2004 report using an older approach. This suggests that at present rates of consumption, we would need 1.39 Earths to insure that future generations are at least as well off as we are now.[2]

A major part of our footprint is the release of greenhouse gases into the atmosphere. Most of these gases are the result of burning fossil fuels. These gases form a type of thermal blanket around the Earth. The increase in emissions of greenhouse gases has contributed to the warming of the Earth and climate change as more heat is trapped in the atmosphere. Climate change is caused by the emission of gases from vehicles, industry, power plants, and deforestation. Carbon dioxide is the primary green house gas. Researchers have also developed tools to help us calculate the amount of carbon dioxide that is released from our lifestyle activities.

I believe it is important for us as North American Christians to reflect and act on the inequity in the greenhouse gas emissions due to our consumptive lifestyles.

The Nature Conservancy has a Web site with a carbon dioxide calculator (http://www.nature.org/initiatives/climatechange/calculator). The amount of carbon dioxide released is measured in tons. The number of tons of carbon dioxide is greater than the weight of the product being burned because two atoms of oxygen (total weight of 32) combine with one atom of carbon (weight of 12) in the burning process. You will find a good list of actions you can take to reduce your carbon dioxide output on The Nature Conservancy Web site.

Other carbon dioxide calculators can also be found on the Internet. Many of the organizations that post these Web sites are selling carbon credits or offsets. By supporting projects that work at reforestation, energy efficiency, and renewable energy, the amount of carbon dioxide in the atmosphere is reduced. An Internet search for "carbon offsets" will yield multiple organizations with carbon dioxide reduction objectives. Mennonite Central Committee has many such projects, such as those found at www.mcc.org/careforcreation/donate.

I believe it is important for us as North American Christians to reflect and act on the inequity in the greenhouse gas emissions due to our consumptive lifestyles. The worldwide average is five tons of carbon dioxide per person per year while the average American emits 22 tons of carbon dioxide every year.

Group questions

1. *Discuss the connection between "knowing the wonderful works of God" and "cycles of life and death, increase and decrease, use and reuse." How does this relationship encourage us to be good stewards of creation?*

2. *Find an ecological footprint calculator on the Internet and complete the quiz questions. What is the average ecological footprint of the group? Do the same with a carbon footprint calculator.*

3. *What are the actions you have taken to reduce your carbon footprint? What is the value of carbon offsets?*

Jesus and the footprint

God has been interested in healthy relationships between humans and all creatures since the beginning of time. "Tilling and keeping" the garden in Genesis 2 is God's way of saying "watch your ecological footprint." Life is lost if the Earth is not kept. God's desire for ongoing ecological well-being is reiterated in the covenantal statements in Genesis 9:9-11 following the flood.

> "As for me, I am establishing my covenant with you and your descendants after you, and with every living creature that is with you, the birds, the domestic animals, and every animal of the earth with you, as many as came out of the ark. I establish my covenant with you, that never again shall all flesh be cut off by the waters of a flood, and never again shall there be a flood to destroy the earth."

In this covenant, God made a commitment not to destroy the Earth. Because of our love for God and our desire to be the image of God, a reciprocal covenant from us is an important act. Lovingly caring and nurturing health within all of God's creation is an outstanding act of stewardship. The books of Moses repeatedly give directions for how people should care for creation – or in present-day terms, mind our ecological footprint. For example, in the book of Deuteronomy instructions are given for respecting trees (20:19), caring for a bird with young (22:6-7), dealing with waste (23:13), and being mindful of the poor (24:19-21). As we have noted earlier, the prophets also called for repentance and justice that included an ecological stewardship.

Lovingly caring and nurturing health within all of God's creation is an outstanding act of stewardship.

Jesus' life and death portray the love of God in ways that supersede the law. While Jesus does not give us a list of ecological principles to live by, he did teach an ethic based on love, integrity, and righteousness. Jesus' actions toward "the least of these" – the poor, women, children, and the vulnerable – demonstrate a kind of footprint we should leave behind. Three areas to consider from Jesus' life and teaching that relate to our ecological footprint are jubilee, restoration, and simple living.

Jesus and Jubilee

Jesus proclaims the Jubilee in new ways during his three years of ministry. The Sabbath and the year of Jubilee were key themes in the writings of Moses and the prophets. Seven days, seven years, and seven times seven years all had great significance for ecological health, both for people, and all of creation. For example, in Leviticus 25:18-19 – a chapter on the year of Jubilee – we read how people and land experience harmony. "You shall observe my statutes and faithfully keep my ordinances, so that you may live on the land securely. The land will yield its fruit, and you will eat your fill and live on it securely."

Nature and people, land and plants, creatures and soil all experience renewal. Jesus proclaimed this same Jubilee at the opening of his ministry (Luke 4:16-19) and in his parables. The parables of the good Samaritan (Luke 10:30-37), the rich fool (Luke 12:16-21), and the rich man and Lazarus (Luke 16:19-31) each address parts of the world out of balance. The issues named relate to rich and poor and the disparity between them. Jesus is calling for the application of Jubilee, which would enable a fresh start for everyone. This reality is emphasized in the results of our ecological footprint.

Jesus and restoration

Restoration grows out of reconciliation. Bringing fullness to a depleted life is a principle taught by Jesus. For example, after Nicodemus questioned the restoring work Jesus was doing, "Jesus answered him, 'Very truly, I tell you, no one can see the kingdom of God without being born from above'." (John 3:3) Regeneration – the renewing of life – has its source in God above. The rebirth is possible only when the God who created the world as an expression of his love takes action. As Jesus declares in John 3:16, God loved the whole world through the process of sending the Son. This is an initial and ongoing way God works in the world. The acts of preserving and renewing are displayed throughout Christ's life, death, and resurrection. These acts are also the basic definition of sustainability. The restoration is for all that God has created. How grand that the word "all" is used! Everything that God has made is included. Our ecological footprint can be altered through restorative acts.

Our ecological footprint can be altered through restorative acts.

Jesus and simple living

Jesus was radical in his teaching about what is needed to maintain life. He repeatedly called for a simplification in lifestyle that is not driven by materialism. When the disciples responded to Christ's call to follow, they abandoned everything (Luke 5:11). At one point, he sent out seventy disciples with no purse, no bag, and no sandals (Luke 10:4). These acts of simple living greatly reduced the disciples' ecological footprint!

These are difficult instructions and directives for us to come to grips with in the midst of our North American context. One of the people who helped shape my thinking about simple living is Doris Janzen Longacre. In *Living More with Less*, she outlines five life standards for living faithfully and simply: (1) do justice, (2) learn from the world community, (3) nurture people, (4) cherish the natural order, and (5) nonconform freely.

Finding direction for the challenge of each standard is found within the teachings of Jesus and within community. The standards, rather than creating a restriction, offer a freedom within the belief that by using less we experience more for our community and ourselves. The actions that emerge from this work are to be practiced with a consistency. Longacre notes that while the initial steps may seem insignificant, the sum of them creates change and health.

> One tiny decision. Nothing that will change the world. But that's the kind which form the building blocks of our lives. It's the sort of decision on which we often falter if we slide unthinkingly in the groove of our society. More-with-less standards don't come naturally right now in North America. Do justice. Cherish the natural order. Nurture people. These and other standards must become second nature for Christians, part of the heredity of our new birth. Many decisions still will be hard. But strong standards rooted in commitment to Christ offer hope for better choices.[3]

Assessing our ecological footprint creates a baseline for knowing where we are in our impact on creation. Reflecting on and being honest with the results leads us to a conviction to change our ways. Renewal occurs as we reduce our consumption and engage in restorative acts. Choosing a Christ-style of living within community will result in a lighter ecological footprint.

Prayer:

Dear Lord, You too walked on this Earth.

We thank you for taking on flesh like ours.

But we struggle with the reality of our walk.

Our North American footprint is large and heavy.

We know there is a connection with our way of living

and the oppression of other people in the world.

We don't want this to be so. Our lives are filled with many good things.

Parting with them is hard, for they give us comfort and security.

Forgive us for having our securities so rooted in things rather than in your love.

We seek repentance. We desire change.

Grant us wisdom to live in your joy by living with less.

May we be participants with you in restoration.

Amen.

End questions ———————————————————————

1. What might 21st century Jubilee look like? What is the North American Christian's role in this?

2. What are the ways in which you have experienced restoration in your life? What are the restorative acts you could take related to your ecological footprint?

3. How does simple living apply to your life? Are there actions you would like to be taking that would lessen your ecological footprint and contribute to ecojustice?

An activity to do at home:

Go to an online ecological footprint calculator and answer all the questions. Note the results. Then repeat the exercise several times, putting in different answers that would represent a change in your lifestyle patterns. What makes the most difference? Which of the changes you entered as answers would you be willing to make?

Sources

1   http://www.earthday.net/footprint

2   Jason Venetoulis and John Talberth, *Ecological Footprint of Nations: 2005 Update*, http://www.ecologicalfootprint.org

3   Doris Janzen Longacre, *Living More with Less* (Scottdale, Pa.: Herald Press, 1980), 17.

Food Systems and Water Supplies

"If your enemies are hungry, give them bread to eat; and if they are thirsty, give them water to drink." – Proverbs 25:21

Two gifts

Two primary necessities for life are water and food – both gifts from God. All living cells require water and energy to survive. Humans have the same needs. The human body is approximately 60 percent water. Eating regularly provides the energy needed for an active life. Throughout Scripture we find references to food and water not only in relation to physical needs but also with a spiritual dimension. Jesus says the faithful servant will give food and drink to the hungry. With food and water being so central to life, it is important we explore the creation care stewardship aspects of these two essentials.

The provision of food

"Give us this day our daily bread." – Matthew 6:11

The fruit room in our basement was a wonderful site to behold at the end of the growing season. The vivid colors of fruits and vegetables in glass jars that lined the shelves were simply refreshing. The garden had yielded its bounty again. The work of all hands in our family made this grand collection possible. There were green beans, red beets, yellow peaches, applesauces of various kinds, purple grape juice, white peaches, red tomato juice, green pickles, and orange carrots. On the floor were buckets of onions, potatoes, and rows of squash. The freezer was filled with additional fruits and vegetables.

Sometimes we would just go down and look in awe at the abundance. The hard work was well-known, but we were connected to this food and were delighted with it. Most of this food came from the garden only 200 feet away. While the money in our pockets was limited, we had a special sense of contentment on each trip to the fruit room.

This does not represent the food systems in North America. Today the average food on dinner tables travels 1,300 miles. The average food footprint of each American is 5.2 acres, compared with the world average of 1.9 acres per person. We are in an unsustainable system. This is a hard concept for us to grasp as North Americans when we visit grocery stores. Every aisle is full of goods with a supply that never seems to diminish. Yes, we know there are people who are hungry, but the system we see does not seem to be broken.

The abundance that surrounds us does not give us the right to consume beyond our needs.

As a church we can be polarized over whether the food system is broken or not. Tensions arise as the fossil fuel-based agriculture is challenged. Many people have been fed by this system. It is the source of income for many people, yet as one considers the future limitations of fossil fuel, there is recognition that this food system is not sustainable. The National Sustainable Agriculture Information Service (http://attra.org) draws the following conclusion.

> Sustainable agriculture is the best chance we have to feed the world. Today's industrial food system not only occupies an exorbitant amount of the biosphere's regenerative capacity, it also degrades the productivity of ecosystems, both natural and farmed, which are the very basis of our food supply. Ecological Footprint accounts, by identifying the ecological constraints of food and other human demands, underline why planning for resource and food security are essential strategies for a socially just and ecologically healthy world.[1]

The global statistics paint a grim food system picture. Consider the following quotes from World Hunger and UNICEF – two leading organizations working to address global food issues.

> Poverty is the principal cause of hunger. The causes of poverty include poor people's lack of resources, an extremely unequal income distribution in the world and within specific countries, conflict, and hunger itself. There are an estimated 1.08 billion poor people in developing countries who live on $1 a day or less. Of these, an estimated 798 million suffer from chronic hunger, which means that their daily intake of calories is insufficient for them to lead active and healthy lives.[2]
>
> Malnutrition is devastating. It plays a part in more than half of all child deaths worldwide. It perpetuates poverty. Malnutrition blunts the intellect and saps the productivity of everyone it touches.[3]

The disparity is unfathomable. As North American Christians, we have difficulty knowing how to respond to a need that feels greater than our capacity. We ask, "What difference does it make on the world hunger issue if I reduce my food consumption patterns?" In many ways, the result is very distant and indirect. The reality is that we are called to be good stewards of all of creation. The abundance that surrounds us does not give us the right to consume beyond our needs. Stewardship of our food is an action we can have control over. The sum of many of us taking the following actions seriously will begin to influence parts of the global food system.

> *As stewards of creation, we have the joyful task of participating in God's food system.*

Stewardship of food:

1. Plant your own garden – Small plots can produce amazing quantities of food that don't need to travel very far. And it's good exercise, too!

2. Buy local food – Purchasing locally grown food reduces the amount of energy needed for transportation and aids the local economy – plus you get to know the grower.

3. Eat foods that are in season – Another benefit of buying locally is that you will eat food in the season when it is grown. The freshness of this food is the best. Expecting all types of foods to be available in all seasons of the year adds a huge energy burden to the food system.

4. Make healthy choices – Many of the products available today are lacking in good nutritional value. Learn what is best for your body, read labels, and purchase accordingly.

5. Avoid preservatives and highly processed foods – The chemicals added to food to make them appear more colorful and fresh are not good for your body. Many foods are also highly processed, which reduces the quality and increases embodied energy.

6. Reduce animal protein intake – Significant amounts of food energy is lost as animals convert plant material into muscle protein. An important way to deal with the limited food energy supply is to rely on a diet that is more plant based.

7. Use leftovers – With good refrigeration, there is no need to throw away leftovers.

8. Compost waste – The process of preparing a meal produces scraps that can be composted. Having a compost pile or a worm bin are good ways of getting the nutrients back in your garden.

9. Pay attention to energy – Cook wisely to conserve energy use in preparation. Use energy-efficient refrigeration and avoid food products that use large amounts of packaging.

10. Assist by giving to local food pantries or banks – These are good ways to practice ecojustice locally.

11. Support international service agencies – Be generous to Mennonite Central Committee or other good international food agencies.

In the story of feeding the five thousand in Luke 9:13-17, Jesus gave the following direction to his disciples: "You give them something to eat." (Luke 9:13) The writer of Proverbs shares this wisdom: "Those who are generous are blessed, for they share their bread with the poor." (22:9) As stewards of creation, we have the joyful task of participating in God's food system.

Group questions

1. Share a story of homegrown, cooked food from your life.

2. What do you consider the advantages and disadvantages of a more localized, sustainable agricultural food system?

3. What "food stewardship" suggestion would you add to the list of 11 items?

Water source

"A generous person will be enriched, and one who gives water will get water." – Proverbs 11:25

What a difference is made on water use when you carry all the water that is needed in your household! This was a regular experience while we served for three years in Grenada, West Indies, with Mennonite Central Committee. I was grateful I could carry two five-gallon buckets at the same time. We used the water for drinking, cooking, bathing, and washing clothes. We could make 10 gallons go a long way.

At first there was no indoor plumbing in the house. But even when the plumbing was installed, the water was intermittent. It was discouraging to take a walk up the mountain and find a broken waterline gushing water into the ground while back at home I was carrying water. In our setting, water was not lacking, but the delivery system was antiquated and poorly constructed. These experiences helped me learn not to take water for granted.

> *Practicing good stewardship of water is an important part of creation care.*

The children of Israel experienced a shortage of water – and they grumbled about it. I think I would have, too. Longing for water can turn into panic, which is only resolved by a refreshing drink. Imagine coming to a body of water and finding it not suitable for drinking. (Most bodies of water are not suitable for drinking today without treatment.) It is amazing to read of the purification God provided through Moses in Exodus 15:23-25.

> "When they came to Marah, they could not drink the waters of Marah, for they were bitter; therefore it was named Marah. So the people grumbled at Moses, saying, 'What shall we drink?' Then he cried out to the LORD, and the LORD showed him a tree; and he threw it into the waters, and the waters became sweet." (NASB)

A good drink of water is desirable and valuable! Yet where I live in northern Indiana, it is seen as a cheap resource. Because we are sitting on top of an extensive and abundant aquifer of water, it makes it hard to realize there are limits to good water. As a friend said, "I could pump water all day, and it wouldn't make any difference." But I can't just let the water run because I am aware of the great shortages in other places in the country and around the world.

Global water shortages and the corresponding lack of sanitation create a grim situation. UNICEF reports the following.

> More than 2.6 billion people – forty percent of the world's population – lack basic sanitation facilities, and over one billion people still use unsafe drinking water sources. Lack of safe water and sanitation is the world's single largest cause of illness. In 2002, 42 percent of households had no toilets, and one in six people had no access to safe water. The toll on children is especially high. About 4,500 children die each day from unsafe water and lack of basic sanitation facilities. Countless others suffer from poor health, diminished productivity, and missed opportunities for education.[4]

This global awareness is a reality check for me as I live in a place with abundant water. So, does it really make any difference whether I conserve water or not? While I cannot realistically transport water from northern Indiana to places in need, I can take actions that represent faithful stewardship of water – a gift from God. My local stewardship of water is an indication of my solidarity with those who are not as fortunate. Water is one of God's gifts, and therefore caring for the gift is the responsibility of each of us. I believe practicing good stewardship of water is an important part of creation care. The following list represents actions we can take.

Stewardship of water:

1. Fix all dripping faucets – A dripping faucet can leak up to 25 gallons per day. The same is true of a toilet that is leaking due to a poor seal. These are easy corrections we can make that will conserve water and save money.

2. Use water-efficient fixtures – By choosing toilets that require less water for flushing, large volumes of water will be conserved. Low-flow showerheads and faucets will reduce water consumption, while still providing plenty of water to accomplish the task. Front-loading washing machines use much less water.

3. Turn off the water – This is a simple water-saving action. For example, don't let the water run aimlessly while you are brushing your teeth or preparing a meal.

4. Limit length of shower times – Long, hot showers are a luxury in North America. By limiting the time in the shower, water is conserved and energy for heating water is saved.

5. Avoid unnecessary washing of clothes – Most clothes don't need to be washed as often as we are prone to wash them. Cleanliness can be maintained with fewer washings – and the clothes last longer.

6. Avoid water-demanding landscaping – Much of the landscaping around homes today requires large amounts of water to maintain the green look. The choice of native trees, bushes, and grasses will reduce the need for watering and fertilization. Less watering also means less fossil fuel for mowing.

7. Use drip irrigation – Some of the most productive food systems in the world use water-efficient drip irrigation. This delivers the water directly to the point of need. As a result, much less water is lost to evaporation.

8. Collect rainwater – A great source of water for our outdoor use is rain. Collecting water in rain barrels or cisterns stores up rainwater to use later. Water pumped from cisterns can also be used for laundry and flushing toilets.

9. Carry drinking water – We have developed a habit of expecting bottled water, an energy and resource wasting practice. Carrying a durable water bottle with water from your tap will save the fossil fuel required for manufacturing the disposable bottles and all the energy of transportation.

10. Pay attention to what happens to your wastewater – Drains and toilets are great, but where does the wastewater go? You should know your system of treating wastewater to ensure good practices are in place. The less water you use, the less water needs to be treated.

11. Pay attention to what you mix in water – Water is a great solvent. We mix all kinds of chemicals in water. The problem is purifying the water once toxins – such as phosphates in soaps and pesticides – have been mixed with water. Without proper purification, this becomes harmful to the next user.

12. Know your watershed – Where your water comes from and flows to is a fascinating study, but we are often unaware. Learning to understand the source of your water will help in forming a better respect for water.

13. Assist in cleaning local creeks and rivers – As we learn to respect our watersheds, we can act on improving their health by joining in caring for these systems locally and regionally.

14. Support international water projects – The global water needs are enormous. Giving financial support to water projects (such as those of Mennonite Central Committee and other global agencies) is an excellent way of sharing from our abundance.

> *Growing in our awareness of the value of water – not just seeing it as a cheap commodity – is an important act of stewardship of creation.*

Growing in our awareness of the value of water – not just seeing it as a cheap commodity – is an important act of stewardship of creation. Jesus guided the woman at the well into a better understanding of water. In John 4:10, Jesus opened her understanding to a new kind of water – living water: "If you knew the gift of God, and who it is that is saying to you, 'Give me a drink,' you would have asked him, and he would have given you living water." We are to practice stewardship of the gifts of water – both the physical and spiritual.

Prayer:

Oh God, we thank you for being the source of life.

Repeatedly through your Word and through your creation,

you have shown us the importance of food and water.

Both of these are gracious gifts. We give you thanks.

You ask us to share food and water with all – even our enemies.

We confess that our sharing is meager in comparison with our abundance.

Teach us to be good stewards of both the gift of food and the gift of water.

We pray for wisdom to know how we can be better partners in sharing these

gifts with all our neighbors near and far. We desire a greater equity.

May you be glorified as we care for these gifts.

Amen.

End questions

1. *Share three facts you know about your watershed. What are needs that you are aware of in your watershed? If you don't know anything about your watershed, what are some action steps you can take to learn about it?*

2. *What are "water conserving" stewardship activities that you can add to the list?*

3. *Reflect on the two gifts of water God has given us – physical water and living water. What are the stewardship connections between these two great gifts?*

An activity to do at home:

Draw a diagram that represents where both your food and water come from. You could use different colors and symbols to indicate sources of food and water. Try to represent the geographical area needed to supply you with your daily food and water.

Sources

1 http://www.attra.org

2 http://www.worldhunger.org

3 http://www.unicef.org/nutrition/index.html

4 http://www.unicef.org/wes/index_31600.html

Buildings and Transportation Systems

"None of this is impossible. Western Europeans use half as much energy per capita as Americans do, and not through different technologies but through different lifestyles and different policies. But none of this is easy." – Bill McKibben[1]

Two essentials

Buildings and transportation are requisites of our daily lives. We spend a great deal of our time indoors. Most of these structures – homes, businesses, schools, and churches – are an architectural combination of rectangular boxes designed to provide shelter and comfort. We travel to and from various "boxes" in "boxes" with four wheels. Almost all of the "boxes" are air conditioned to make sure we are not too cold or too hot. Typically the windows in these "boxes" are kept closed, which makes the conditioning more efficient. There is a strong connection between cars and buildings in much of North America. Many buildings are designed with the car in mind.

Our earlier lessons helped us understand we are stewards of all things. Our call to be earthkeepers comes from our theology and ethics. We are also cognizant that all we do contributes to an ecological footprint – and the data shows our patterns of consumption in North America do not promote ecojustice. It is in this context that we briefly study the ecological stewardship issues of buildings and transportation.

Our shelter

"We send subtle but powerful messages about our worldview in the way that we design our buildings." – Bill Browning, Rocky Mountain Institute, 1998.

Watching beavers swim and work is one of the pleasures of canoeing in the Boundary Waters of northern Minnesota. Seeing them swim under the canoe in crystal clear water is a delightful moment. If you sit quietly, you can observe them gnawing on branches or cutting down a tree. They swim with pieces of wood in preparation for building a lodge or creating a dam. Homes and places are important to them, just as is true for all creatures.

Houses and buildings are also needed in our lives. They provide shelter, comfort, and a place for effective work. Buildings also add a sense of permanence, as well as a setting for memories. It is good for society when buildings are structurally sound and function well for a long time. A number of stewardship and sustainability concepts emerge from this perspective. (1) Choose the best materials for durability and low environmental impact. (2) Design the building for energy efficiency. (3) Determine the spaces for human health. (4) Think about the long-term potential uses for the building.

Building green is a good way to practice stewardship of the Earth.

Buildings place high demand on the Earth's resources. The U.S. Green Building Council, a nonprofit organization committed to promoting sustainability in all types of buildings, notes the following list of ways that buildings impact the environment and our resources.

> In the United States alone, buildings account for:
> 65 percent of electricity consumption,
> 36 percent of energy use,
> 30 percent of greenhouse gas emissions,
> 30 percent of raw materials use,
> 30 percent of waste output (136 million tons annually), and
> 12 percent of potable water consumption.[2]

I believe building green is a good way to practice stewardship of the Earth. It is often argued that it costs too much to build with a sustainable design. Using full-cost accounting – where construction costs, operating costs, and environmental impacts are all calculated – demonstrates that building green is the most cost-effective way. Alex Wilson, a leader in sustainable design, names six benefits of building green.

1. First-cost savings by reducing infrastructure costs, reducing material use, savings related to construction waste disposal, potential tax credits, and other incentives.

2. Reduced operating costs through lower energy costs, lower water costs, greater durability and fewer repairs, reduced cleaning and maintenance, lower insurance costs, and reduced waste generation within the building.

3. Other economic benefits include increased property value, more rapid sales of homes and condominiums, easier employee recruiting, reduced employee turnover, reduced liability risk, and a positive public image.

4. Health and productivity benefits include improved health, enhanced comfort, reduced absenteeism, improved worker productivity, improved learning, and increased retail sales.

5. Community benefits include reduced demand on municipal services, reduced erosion and stormwater runoff, reduced automobile use and traffic congestion, and sprawl and support of local agriculture.

6. Environmental benefits occur because of reduced global warming impacts, minimized ozone depletion, reduced resource extraction impacts, reduced toxic emissions, reduced energy and other impacts of transporting materials, reduced contributions to local and regional air pollution, reduced local and regional water pollution, protection of biodiversity, and increased environmental awareness.[3]

Each of the above points deserves discussion, but space is limited within this study. A key point is to take the impact of buildings seriously, recognizing that choices, wise or unwise, have a long-term effect. An important question for designing all types of buildings is, "How big a space do we really need?" Sarah Susanka, a home design architect, comments on sizes of homes.

> The inspiration for The Not So Big House came from a growing awareness that new houses were getting bigger and bigger but with little redeeming design merit. The problem is that comfort has almost nothing to do with how big a space is. It is attained, rather, by tailoring our houses to fit the way we really live, and to the scale and proportions of our human form.[4]

An excellent way to be more sustainable is to use existing buildings. While existing houses have their limitations, they represent a lesser impact on the Earth. The resources were already put together back at the construction period of the house. The main issue to consider is how to make its current operation more efficient. Here is a list of some of the energy-saving practices one can do with an older home:

1. Weatherize cracks, door seals, and windows.

2. Lower the thermostat setting and wear layers of clothing during the winter.

3. Raise the thermostat setting during hot weather.

4. Install a programmable thermostat to automatically regulate temperatures.

5. Add insulation to your ceilings and walls.

6. Close off parts of the house you do not use on a daily basis.

7. Invest in energy-efficient heating and cooling systems.

8. Replace old windows with more energy-efficient models.

9. Disconnect electronic equipment that is not in use.

10. Turn down the thermostat in your water heater.

A key point is to take the impact of buildings seriously, recognizing that choices, wise or unwise, have a long-term effect.

Careful planning is essential because there are long-term results based on the choices we make. This principle is taught in the construction stories in the Bible. Noah followed all the directions God gave him while building the ark (Genesis 6:22). Great detail was given for building the tabernacle, which many people worked at accomplishing (Exodus 25-31 and 36-39). David recorded all the plans for building the temple, which Solomon completed (1 Chronicles 28:11). The importance of careful design and construction in these "places for God" is an indication of the attention we should give to buildings today.

Jesus teaches us to count the costs, both in spiritual commitments and in planning. Perhaps in Luke 14:28, he also meant full-cost accounting as well as first-cost! "For which of you, intending to build a tower, does not first sit down and estimate the cost, to see whether he has enough to complete it?"

Group questions

1. *What are the most important values you see in building green?*

2. *What are the energy-saving efforts you have completed in your home? Your church? Your business?*

3. *Read the account of the careful planning that was done for the temple in 1 Chronicles 28. What are the good building principles that can be learned from this story?*

Getting from here to there

"More than anything else, fossil fuel has allowed us to stop being neighbors to each other, both literally – we move ever farther into ever emptier suburbs – and figuratively – we depend less and less on each other for anything real. (The SUV, with its almost invariably single passenger, is the symbol of this trend.)" – Bill McKibben[5]

The larger creatures of the Earth move to areas beyond their homes by using feet, tails, wings, arms, bellies, and fins. Tiny creatures have fascinating ways of transporting themselves from one spot to another – from flagella to attachment or floating to amoebic undulations. Most of these movements relate to collecting food and finding mates. Humans are wonderfully made as well with arms and legs. While these are quite effective in many activities, we also have a thinking capacity to design transportation devices and systems.

During Christ's lifetime, transportation was limited to walking, riding on an animal, or using a boat. All of these represent sustainable models, but feel restrictive for our lives in the 21st century. So what does Jesus have to say to us about being good stewards of creation through our choices in transportation?

In 2002, the Evangelical Environmental Network designed an awareness program (http://www.whatwouldjesusdrive.org) around the question, "What would Jesus drive?" Just because driving a vehicle was not a choice for Jesus does not make this an irrelevant question. The ecological impact of present-day vehicles – powered by fossil fuels – is great. Human wellness is compromised, resources for the poor are depleted, and creatures lose health and habitat. The life of Christ instructs us in the meaning of holistic simplicity. Jesus teaches us to care for the poor and the vulnerable. The impact caused by our transportation choices raises moral questions we as Christians need to face together.

The impact caused by our transportation choices raises moral questions we as Christians need to face together.

Travel patterns in North America revolve around the automobile. It represents a gripping challenge. We count on our cars for so many parts of our lives. Lester Brown, founder of the Earth Policy Institute, highlights this problem – a problem we find hard to solve even in the face of the moral questions it raises.

> There are many reasons to question the goal of building auto-centered transportation systems everywhere, including climate change, air pollution, and traffic congestion. But the loss of cropland alone is sufficient. Future security now depends on restructuring transportation budgets – investing less in highway infrastructure and more in a land-efficient rail, bus, and bicycle infrastructure.[6]

Transportation choices represent the incongruity found in my commitments to caring for creation. On the one hand, I walk 50 steps to my garden to gather wonderful homegrown vegetables. It is hard to beat that food and transportation system. But on the other hand, I drive 280 miles per week to be involved in the "good work" of environmental education. What makes my choice even worse is the fact that my work takes me away from the office on an irregular schedule, which means I am not available to carpool with other staff members who live close.

The car can feel like a gift to my life, for it meets my demand for convenience, flexibility in my schedule, and serving my wide-ranging interests and opportunities. Plus it is possible because of "cheap" fossil fuel. The low cost is reality even when the price at the pump is $3.15 per gallon as I write this lesson. The high cost comes with the depletion of resources in manufacturing and maintaining the car, the inequity that this transportation choice represents, and the wide-ranging impact of pollution. While I can pat myself on the back for driving a fuel-efficient compact car, I still contribute almost 150 pounds of carbon dioxide to the atmosphere each week.

The release of greenhouse gases from transportation is one of the major issues for creation care stewardship to address. The following list shows the number of pounds of carbon dioxide released per person per mile of travel. It is also important to remember that carbon dioxide is only one of the pollutant gasses released as we burn fossil fuels.

CO_2 pounds per passenger per mile[7]

Bike or walk	0.00
Mass transit (3/4 full)	.26
Carpool (3 people, 21.5 mpg)	.37
Intercity train (U.S. average occupancy)	.45
Economy car (solo driver, 40 mpg)	.59
Mass transit (1/4 full)	.75
Jet (U.S. average occupancy)	.97
Average car (solo driver, 21.5 mpg)	1.10
Sport utility vehicle (solo driver, 15 mpg)	1.57

The release of greenhouse gases from transportation is one of the major issues for creation care stewardship to address.

My transportation dilemma goes on because of air travel. I don't fly frequently due to work assignments, but we have three children who live in Oregon, Alabama, and California – all a long way from northern Indiana. Flying helps us have more time with our children and less time in travel. It is usually less expensive – if you only count the ticket dollars and not the ecological impact – than other forms of transportation. My transportation patterns make the results of an ecological footprint quiz horrendous. If everyone in the world lived like me, it would take 3-4 more earths.

The chart below shows the changes in expenditures that have occurred in U.S. households over an 85-year period. Housing and transportation have become the two largest costs, with the more significant increase in transportation. In light of our previous study, it is interesting to note the 28 percent decrease in food expenditures. While the percentage costs of transportation dipped in the 2005 report, the costs have accelerated in the years since. More and more of our personal resources are going into a category that is also harming the Earth and all creatures.

Average Household Expenditures (U.S.)

Year	1919[8]	1950[8]	1987[8]	2005[9]
Food	41%	32%	19%	13%
Housing	27%	26%	34%	33%
Transportation	3%	14%	26%	18%
Clothing	18%	12%	5%	4%
Health care	5%	5%	4%	6%

A considerable change is needed in our transportation systems to help bring restoration and hope to all of us as keepers of the Earth. Bill McKibben, a writer who addresses environmental issues out of his stewardship commitment as a Christian, writes,

> Not all the answers are technological, of course – maybe not even most of them. Many of the paths to stabilization run straight through our daily lives, and in every case they will demand difficult changes. Air travel is one of the fastest growing sources of carbon emissions around the world, for instance, but even many of us who are noble about changing light bulbs and happy to drive hybrid cars chafe at the thought of not jetting around the country or the world. By now we're used to ordering take-out food from every corner of the world every night of our lives – according to one study, the average bite of food has traveled nearly 1,500 miles (2,414 kilometers) before it reaches an American's lips, which means it's been marinated in (crude) oil. We drive alone because it's more convenient than adjusting our schedules for public transit. We build ever bigger homes even as our family sizes shrink, and we watch ever bigger TVs, and – well, enough said. We need to figure out how to change those habits.[10]

What a dilemma we face as North Americans in transportation! Our acts cause suffering for others, even as our life is mostly one of comfort. We wrestle with the issues in order to make better choices that will aid in restoring wholeness. Read Romans 5:3-5 and consider the spiritual pattern laid out here of a hope-filled outcome as we struggle to be faithful stewards.

> "... knowing that suffering produces endurance, and endurance produces character, and character produces hope, and hope does not disappoint us, because God's love has been poured into our hearts through the Holy Spirit that has been given to us."

Prayer:

Oh Lord,

We give you praise for being our shelter and for carrying us on your wings.

You have provided for our spiritual well-being

and have filled our lives with good things.

For these we give you thanks.

Yet we are perplexed, challenged, and sorrowful by how

our ways of constructing buildings and transportation systems

have drained and damaged so many resources on the Earth.

Guide us in finding ways to improve our designs

and choices in buildings and travel.

Your wisdom is able to help us find ways that are

more efficient, effective, and just.

May our stewardship actions be to your honor and glory.

Amen.

End questions

1. What are the choices you have made – or would like to make – to reduce your carbon footprint related to travel?

2. How does the Bill McKibben quote at the end of this section represent life as you know it today? How might you make it better fit your worldview?

3. How do you see the model for producing hope in Romans 5:3-5 apply to solving the dilemmas of inequitable transportation systems?

An activity to do at home:

Calculate your household's annual costs for housing and transportation. How does this compare with the national data? How might you lessen your ecological footprint in these two categories?

Sources

1   Bill McKibben, "Hot and Bothered: Facing Up to Global Warming," in *The Christian Century Magazine*, July 11, 2006.

2   U.S. Green Building Council, http://www.usgbc.org/DisplayPage.aspx?CMSPageID=1718.

3   Alex Wilson, "Making the Case for Green Building," in *Environmental Building News*, April 2005, www.buildinggreen.com

4   http://www.notsobighouse.com

5   Bill McKibben, "Hot and Bothered: Facing Up to Global Warming," in *The Christian Century* Magazine, July 11, 2006.

6   Lester R. Brown, *Plan B: Rescuing a Planet under Stress and a Civilization in Trouble* (New York: W. W. Norton and Company, 2003), 147.

7   http://www.sightline.org

8   David S. Johnson, John M. Rogers and Lucilla Tan, "A Century of Family Budgets in the United States," *Monthly Labor Review* (www.bls.gov/opub/mlr/2001/05/art3full.pdf), May 2001, 28-46.

9   "Consumer Expenditures in 2005," U.S. Department of Labor, U.S. Bureau of Labor Statistics, Report 998, February 2007, http://www.bls.gov/cex/csxann05.pdf, 3.

10 Bill McKibben, http://magma.nationalgeographic.com/ngm/2007-10/carbon-crisis/carbon-crisis-p3.html.

Climate Change and Global Dynamics

"Neglect in protecting our heritage of natural resources could prove extremely harmful for the human race and for all species that share common space on planet earth. Indeed, there are many lessons in human history which provide adequate warning about the chaos and destruction that could take place if we remain guilty of myopic indifference to the progressive erosion and decline of nature's resources." — R. K. Pachauri, Chairman of Intergovernmental Panel on Climate Change[1]

Climate change

The repeated news headlines pertaining to climate change are daunting and frequently hopeless. We are faced with several responses: choosing denial, falling into despair, or seeking change. How do we engage such an enormous challenge that threatens the core of our existence?

I enjoy walks that give me a view of the sky. The walk may be at night or in the day, in open fields or along city streets, with a view of the horizon or limited by tree canopy, clouded or brilliant. Seeing the firmament speaks to my spirit and reminds me of my place in the universe. The sky is literally and figuratively beyond me. The psalmist reminds us of our role in relationship to creation in Psalm 89:11: "The heavens are yours, the earth also is yours; the world and all that is in it – you have founded them."

How could I – and my fellow inhabitants – actually change the climate of this place?

The Earth – in fact the cosmos – belongs to God. As we have noted previously, we are to be keepers of this place as faithful stewards. On the one hand, I accept God's ownership of creation and am committed to my stewardship role. On the other hand, from my limited view of the sky, I find it difficult to comprehend how much we have impacted the Earth and its systems. As I gaze across the expanse of the sky, I feel so small. The enormity of the atmospheric ocean is mind-boggling, let alone considering the magnitude of the universe. So how could I – and my fellow inhabitants – actually change the climate of this place?

The weather changes all the time. We see time marked by seasons along with its corresponding weather. Global climate change is another matter. In this case, the patterns of the Earth's climate begin to vary beyond the typical range of fluctuations. This includes rising temperatures in the atmosphere and in the oceans, resulting in the melting of glaciers and polar ice and increasing sea levels and areas of desert.

Determining whether change has occurred requires a large amount of data collected over a long period, all of which is studied carefully. The best information available on climate change comes from the Intergovernmental Panel on Climate Change (IPCC) – the recipients of the 2007 Nobel Peace Prize. The IPCC is an international organization that has relied on more than 2,500 scientific expert reviewers and authors from more than 130 countries. Their fourth assessment report issued in November 2007 states,

> Warming of the climate system is unequivocal, as is now evident from observations of increases in global average air and ocean temperatures, widespread melting of snow and ice, and rising global average sea level.[2]

The message in this, as well as earlier reports, needs to be heeded. While it has received international acclaim, it is also important for us as Christians to take note of the report's meaning. In 2006, 86 evangelical leaders signed the Evangelical Climate Initiative statement. The leaders carefully considered the information and interacted with trusted scientists. The following statement comes from the preamble of the ECI statement.

> Over the last several years many of us have engaged in study, reflection, and prayer related to the issue of climate change (often called "global warming"). For most of us, until recently this has not been treated as a pressing issue or major priority. Indeed, many of us have required considerable convincing before becoming persuaded that climate change is a real problem and that it ought to matter to us as Christians. But now we have seen and heard enough to offer the following moral argument related to the matter of human-induced climate change.[3]

The remainder of the statement articulates the reality of climate change, the importance for Christians to take seriously the resulting impact, and to act in ways that will reduce the change. The recognition of climate change and its impact on all Earth's residents is particularly evident among the poor in the world. The Evangelical Climate Initiative states,

> Poor nations and poor individuals have fewer resources available to cope with major challenges and threats. The consequences of global warming will therefore hit the poor the hardest, in part because those areas likely to be significantly affected first are in the poorest regions of the world. Millions of people could die in this century because of climate change, most of them our poorest global neighbors.[4]

Accepting our moral responsibility in this global perspective is critical.

Accepting our moral responsibility in this global perspective is critical. Even with this looming evidence, it is easy to question whether our human actions have caused this dilemma. While we do not want to shirk our responsibility to care for the poor, we struggle with accepting the blame for the root cause. There are multiple factors that can cause climate change, but the leading factor contributing to the present reality is human action. Consider the statement from the IPPC report.

> Global GHG (green house gas) emissions due to human activities have grown since pre-industrial times, with an increase of 70 percent between 1970 and 2004. There is very high confidence that the net effect of human activities since 1750 has been one of warming.[5]

The primary culprit leading to global warming is the emission of green house gases (GHG), most of which is due to the burning of fossil fuels. The greenhouse effect is important for life on Earth. Radiation from the sun enters through our atmosphere, warming the Earth. The gases in the atmosphere trap the energy, which keeps the Earth from becoming too cold at night, thus sustaining life. The problem arises when we burn fossil fuels. More and more greenhouse gasses are added to the atmosphere – especially carbon dioxide – preventing less and less of the sun's energy from escaping the Earth. The result is the warming of the Earth's surface and air, which leads to long-term changes in the climate.

There are multiple factors that can cause climate change, but the leading factor contributing to the present reality is human action.

One of the inequities in this dilemma is the wealthy of the world have access to the largest amounts of fossil fuel and the people most effected by climate change are the poor of the world. The burning of coal to produce electricity for homes and industry and the use of oil to power our transportation systems are privileges that represent the lack of balance between the industrialized and nonindustrial worlds. The United States produces one-fourth of all the world's greenhouse gases while having only five percent of the world's population. The injustices are unsettling. Prophetic statements often are. Consider the message spoken by the prophet Isaiah to the king of Babylon.

> "Those who see you will stare at you,
>   and ponder over you:
> 'Is this the man who made the earth tremble,
>   who shook kingdoms,
> who made the world like a desert
>   and overthrew its cities,
>   who would not let his prisoners go home?' " (Isaiah 14:16-17)

We as North Americans are guilty of being like this king in many ways. Our consumer-driven materialism has caused the Earth to tremble. Our methods of extracting the Earth's resources have made parts of the world like a desert. Wars are fought over the supplies of oil, resulting in cities being overthrown. The poor of the world are trapped by our domination of the resources.

Christians have a message that can overcome the hopelessness and denial found in many places today. It is the message of reconciliation through Christ.

Christians have a message that can overcome the hopelessness and denial found in many places today. It is the message of reconciliation through Christ. We are called to be reconcilers with Christ. The good news of Christ is to be applied to the human soul, but also to all aspects of creation. While human actions have caused the climate to change, a corrective course can also be taken. The IPCC report affirms the potential for limiting climate change.

> Many impacts can be reduced, delayed, or avoided by mitigation. Mitigation efforts and investments over the next two to three decades will have a large impact on opportunities to achieve lower stabilization levels. Delayed emission reductions significantly constrain the opportunities to achieve lower stabilization levels and increase the risk of more severe climate change impacts.[6]

Restorative action is our call. There is resilience in many of Earth's systems, which is to our advantage in restoration. But we have to be active in that process. Without the functions and processes of nature, we can't restore, but working alongside these processes, we can make positive change occur. It is important to find ways to reduce our carbon emissions (see chapters 7-9). Alternative energy sources are critical, such as wind and solar power. Secular writers, such as Lester Brown, also appeal for a change in our mode of operation.

> Plan B is the only viable option simply because Plan A, continuing with business as usual, offers an unacceptable outcome – continuing environmental degradation and disruption and a bursting of the economic bubble.[7]

Since we as humans have adversely impacted and degraded the Earth's systems one action at a time, we should also seek restoration one action at a time. Our individual and collaborative work will make a difference. It is part of the good news message.

Restorative action is our call.

Group questions

1. Share an article from the news of the past week that referred to climate change or global warming. What are the kinds of topics being addressed?

2. How important is scientific information in determining your own thinking and course of action related to climate change?

3. What potential do you see for Christians working together in response to climate change?

Global dynamics

"But religious people need to have confidence in the power and truth of their own sacred texts. Only to the extent that religious people take seriously their teaching – about the inherent dignity of all people, the need to treat others as we would be treated, and increasingly, the need to care for the natural environment – will the power of those teaching be released." – Gary Gardner[8]

There is always a stir of excitement upon spotting a bird fluttering in a mist net on a dewy summer morning. I saw a flash of yellow in the net as I walked alongside a student researcher. Gently the student grasped the bird in her hand, untangled it from the net, and placed it in a soft little bag to carry it back to the banding station. Once at the station, the bird was taken out of the bag for identification, data collection, and placing a small numbered band on a leg. This was a yellow-breasted chat. What a delight to look into its eyes and examine its plumage up close.

The question that came to my mind was, "Where is its home?" I knew this bird's nest was in the brush not too far from the net. While I didn't see the nest, I knew it was close because of the small territory that makes up the nesting habitat of the yellow-breasted chat. But Merry Lea Environmental Learning Center in northern Indiana, isn't its only home. The chat migrates to Central America – perhaps Belize – in the winter. This small bird has two homes spanning thousands   of miles across the globe. Caring for both homes of the yellow-breasted chat is critical. It cannot survive if either home place is lost or damaged. Deforestation or intensive farming could impact either home.

The dynamics of human impact globally are sobering. Homes are lost for many people – creatures of all kinds. The following six examples are brief snapshots that represent a long list of the assaults on the global environment in which we live.

The dynamics of human impact globally are sobering.

1. Air pollution – This environmental damage is estimated to cause approximately 2 million premature deaths worldwide per year. More than half of this burden is borne by people in developing countries.[9]

2. Deforestation – From 1990 to 2005 the net forest loss globally is 18 million acres (28,000 sq. miles) per year.[10]

3. Desertification – As climate change translates into more intense storms, flooding, heat waves, and droughts, more and more communities will likely be affected. Desertification, for example, puts some 135 million people worldwide at risk of becoming environmental refugees.[11]

4. Loss of diversity – Changes in biodiversity due to human activities were more rapid in the past 50 years than at any time in human history, and the drivers of change are not decreasing.[12]

5. Rising water levels – More than 25 percent of Africans live within 100 km of a seacoast. Sea-level rise could significantly increase the number of Africans impacted by coastal flooding, with a worst case being 70 million in this century.[13]

6. Land – Average annual rate of agricultural land converted to developed uses in the United States is 1,234,560 acres.[14]

The plight of the Earth and its inhabitants is grim. I know these facts to be our global reality, yet my day-to-day life is quite comfortable. I do not see most of these predicaments firsthand, which can lead to complacency. The prophet Isaiah again sounds the poetic clarion. He sees a coming judgment that brings to mind our present global quandary.

> "The earth dries up and withers,
>   the world languishes and withers;
>   the heavens languish together with the earth.
> The earth lies polluted
>   under its inhabitants;
> for they have transgressed laws,
>   violated the statutes,
>   broken the everlasting covenant." (Isaiah 24:4-5)

The chapter begins with the impending laying waste of the Earth, which is named as God's action against those who have acted unjustly. I believe this is speaking to the consequences of self-centered living. The balance in the ecological and spiritual designs will be broken when humans treat them with disregard. Ultimately in this passage, God is not the destroyer but rather the one who makes things right. God is the restorer. The destruction is the result of human action.

Much of the disruption in the global dynamic today may be attributed to greed. We have a large amount of "stuff," and we want more. Our mechanized worldview thrives on product. Materialism is an idol that has our attention. We translate it into comfort. The fire of materialism is not satisfied. Fossil fuels feed the machine – and contribute to the damage of the system at the same time. Chris Goodall in *How to Live a Low-carbon Life,* observes the following spiral in the global economy.

> Fossil fuel energy is so cheap and so convenient that its use permeates every aspect of our lives. And as more and more of the world's people move into the market economy, they will want to replace their labour with petrol or electricity.[15]

Much of the disruption in the global dynamic today may be attributed to greed.

Where does the downward spiral stop? We can find principles that help answer this question in Jesus' conversations with two different leaders during his earthly ministry. The first is with Nicodemus, a Pharisee. In the well-known passage from John 3, Jesus talks about the love God has for the cosmos. The statement in verse

19 calls for an examination of our deeds: "And this is the judgment, that the light has come into the world, and people loved darkness rather than light because their deeds were evil."

Jesus' incarnation – the taking on of human flesh – is the light by which to stop the spiral. This light is represented in the sum of his teachings – and in his death and resurrection. We need to choose the light over the darkness.

The second conversation is with an unnamed lawyer in Luke 10. The lawyer was interested in testing Jesus. The exchange in verses 25-28 centered on following the Mosaic law. Jesus said there is life when the law is followed completely. The lawyer may have been better off if he had stopped the conversation at that point, but he asked one more question, "And who is my neighbor?" (Luke 10:29).

This really is the right question to ask today in responding to the global challenges. The whole world has become our neighbor in ways we could not have imagined a century ago. Our pollution travels with the wind, paying attention to no border. Our desire for more – underwritten with cheap fossil fuel – extracts and transports goods from around the world. Jesus gave an answer that made the lawyer squirm. We know it as the Good Samaritan story. The correct reply to the question is found in Luke 10:37: "He said, 'The one who showed him mercy.' Jesus said to him, 'Go and do likewise'."

Our neighbors are found all around the globe, as well as those in the coming generations. This calls for renewed understandings of the concept of being a global citizen. Jesus asks us to show mercy to our neighbors. This cannot be done effectively when greed is the governor of our North American lifestyle.

The world God brought into being was filled with peace, which is also God's continued hope for it. Christ modeled this peace in his teachings and earthly life. He became the ultimate reconciler of all things through his death and resurrection. We are to follow in his footsteps. Our Christian way of living as passionate stewards of the Earth will bring peace and hope. The new life in Christ will be the basis for a global transformation. In conclusion, consider how Lester Brown's hopeful plea for a new type of global economy echoes the teachings of Jesus.

> We can build an economy that does not destroy its natural support systems, a global community where the basic needs of all the earth's people are satisfied, and a world that will allow us to think of ourselves as civilized. This is entirely doable.[16]

Jesus offers an even more abundant hope.

Prayer:

Dear God, You are the lover of the cosmos.

Your love, your patterns, your forgiveness, and your complete peace cannot be fathomed, yet they are sustaining.

This assurance comforts us, for we live in a broken world.

The changes in the climate are vivid reminders of our impact on creation.

We confess that our consumptive lifestyle caused this decay.

Our practices had kept us from extending neighborliness around the globe.

We are grateful for the gift of Christ who is our peace.

Place the cloak of Christ on us so we can extend His peace around the world.

Amen.

End questions

1. What snapshots of global environmental degradation do you know about? How might these dilemmas be resolved?

2. Read John 3:16-21. The word "world" in this passage means cosmos. How does this understanding influence the way we interpret the passage?

3. Reflect on the Good Samaritan story again. What does it mean for you to show mercy to neighbors in the global context?

An activity to do at home:

> *Take a series of walks that focus on viewing the sky. Journal your reflections on what you see and how these walks give insight into the global dynamics discussed in this lesson.*

Sources

1   Acceptance Speech for the Nobel Peace Prize Awarded to the Intergovernmental Panel on Climate Change (IPCC). Delivered by R. K. Pachauri, Chairman, IPCC Oslo 10 December 2007.

2   Summary for Policymakers of the Synthesis Report of the IPCC Fourth Assessment Report DRAFT COPY 16 NOVEMBER 2007, 1.

3   Statement of the Evangelical Climate Initiative, 2006.
     http://www.evangelicalclimateinitiative.org/statement

4   Statement of the Evangelical Climate Initiative, 2006.
     http://www.evangelicalclimateinitiative.org/statement

5   Summary for Policymakers of the Synthesis Report of the IPCC Fourth Assessment Report DRAFT COPY 16 NOVEMBER 2007, 4.

6   Summary for Policymakers of the Synthesis Report of the IPCC Fourth Assessment Report DRAFT COPY 16 NOVEMBER 2007, 20.

7   Lester Brown, *Plan B: Rescuing a Planet under Stress and a Civilization in Trouble* (New York: W. W. Norton and Company, 2003), 199.

8   Gary Gardner, *Inspiring Progress: Religions' Contribution to Sustainable Development* (New York: W. W. Norton and Company, 2006), 166.

9   http://www.who.int/mediacentre/news/releases/2006/pr52/en/index.html

10   http://www.fao.org

11   http://www.worldwatch.org/node/116

12   http://www.millenniumassessment.org

13   IPCC, Impacts, p. 515.

14   http://www.farmlandinfo.org/agricultural_statistics

15   Chris Goodall, *How to Live a Low-carbon Life: The Individual's Guide to Stopping Climate Change* (London: Earthscan, 2007), 13.

16   Lester Brown, *Plan B: Rescuing a Planet under Stress and a Civilization in Trouble* (New York: W. W. Norton and Company, 2003), 221-222.

Congregational Creation Care

11

> *"A parish is a corner of God's creation, a small enclosure within the garden of the earth. Within this boundary we have a responsibility for 'the cure of souls,' not just of the gathered congregation but of all who live within the parish boundary."*
> — James Jones, Bishop of Liverpool[1]

Worship and windows

The windows – and the preacher – in the small country church I attended were instrumental in worship experiences throughout my childhood. The building was a small one-room schoolhouse in northern Minnesota with large windows all along the east side. My father was the pastor, but was also a gentle dairy farmer. What I did not realize is that I was immersed in a church setting that was both eco-logically and holistically invigorating. At that time, my vocabulary did not include words like these to describe my church. I am grateful this setting provided a rich opportunity to learn about God in multifaceted ways.

The biblically based messages my father delivered raised my awareness of God's love for the world and for me. His hope-filled sermons were composed during small blocks of time he managed to take from the full schedule of caring for our small dairy farm. I knew the man who stood behind the pulpit treated all the animals with gentleness and managed the land in ways that brought health to the soil. How wonderful to have harmony between the spoken and lived messages.

> *I heard two messages – one from the pulpit and one from the windows.*

But the wall of windows was also an important element of worship. Not only did they provide good natural lighting, they also offered an ecological view of the world. While listening to the sermon, I had a grand view of all the seasons in the woods right outside the church. There were the pale green colors and frog sounds

143

in the spring, the buzz of mosquitoes and the power of thunderstorms in the summer, the vibrant reds and yellows of fall, and the blankets of snow in the winter. I heard two messages – one from the pulpit and one from the windows – and they were not in conflict with each other.

While the phrase "stewardship of creation" wasn't used in my childhood, it was practiced in daily life – seven days a week. My commitment to creation care was cultivated through the interweaving of spiritual and ecological teachings in my childhood. In this northern Minnesota setting, I learned the truth of the following statement: "We believe that everything belongs to God, who calls us as the church to live as faithful stewards of all that God has entrusted to us."[2]

> *God has placed us, and our congregations, in wonderful ecosystems with an assigned responsibility of faithful stewardship.*

This concise statement combines scriptural declarations on our individual and congregational roles in relationship to the Earth. Consider passages that illustrate this belief statement. First, stewardship for believers begins with acknowledging that God is the creator and owner of all things.

> "The earth is the LORD's and all that is in it,
>   the world, and those who live in it;
> for he has founded it on the seas,
>   and established it on the rivers." (Psalm 24:1-2)

These verses challenge the materialistic tendencies we have in North America. Our drive to have more – own more – is amended when we live in awareness that all things belong to God. Living this understanding helps us turn from an "Earth taking" mode to one of "Earth keeping." God has placed us, and our congregations, in wonderful ecosystems with an assigned responsibility of faithful stewardship. In 1 Corinthians 4:1-2, the concept of stewardship is developed further: "Think of us in this way, as servants of Christ and stewards of God's mysteries. Moreover, it is required of stewards that they be found trustworthy."

This is a beautiful passage, which reminds us of the multiple mysteries of God. They include not only the mystery of salvation, but also all the mysteries of creation. Think of all the wonders in nature that baffle our understanding and stimulate our imaginations! It is important that we have an awareness of the mysteries of God. This is what I experienced in my church setting as a child. Note that this passage instructs us to be stewards of all God's mysteries. But it is not enough that we are named stewards. Our actions as stewards are to be honest, reliable, and honorable – they are to be found trustworthy.

Our actions as stewards are to be honest, reliable, and honorable – they are to be found trustworthy.

The statement that "everything belongs to God" distinctly expresses that stewardship of creation is a call of the church. A third passage I find helpful in illuminating this understanding is 1 Peter 4:10: "Like good stewards of the manifold grace of God, serve one another with whatever gift each of you has received."

God has given each of us gifts – not for the promoting of our self-interests, but for serving. Our church communities are places where these gifts are affirmed, nurtured, and commissioned for action in the world. These gifts include the ability to be good stewards of all things.

Too often we have followed a pattern of discerning gifts that mainly help with the internal functions of the church and evangelism. While utilizing our gifts within these bounds is critical and vital, there is a more complete set of gifts that assist in holistic stewardship of the full gospel of Christ. This passage reminds us of our stewardship role with "the manifold grace of God." As noted from the 1 Corinthians passage, this includes the grace (gift) of salvation and creation. The church is called to serve all the graces of God with our gifts.

The church is Christ's kingdom here on Earth. There are many gifts and many stewards within the church, which is what makes the church vibrant in bearing witness to the good news of the kingdom. While the New Testament lists gifts in several places, I appreciate the added insights about gifts found in the Old Testament. For example, at the end of King David's life, a multitude of gifts were listed that helped sustain all activities in the kingdom. Some of the talents named in 1 Chronicles 23-27 are gatekeepers, musicians, priests, artists, treasurers, overseers, counselors, storekeepers, tillers of the soil, managers of orchards, and keepers of flocks. In 1 Chronicles 27:31 it says, "All these were stewards of King David's property." This is holistic kingdom stewardship – a model for our spiritual and ecological responsibilities.

Our worship services are more complete when we include the intersecting understandings of the spiritual and the ecological.

Becoming more attuned to the church's role of being stewards of all God's mysteries is an important task for us to pursue in our congregations. Our worship services are more complete when we include the intersecting understandings of the spiritual and the ecological. Worship should have a balanced focus on God – the Creator, Redeemer, and Sustainer. Our worship is enhanced when the songs, Scripture, and reflections acknowledge the beauty and wonder of all of creation. Times of confession should include acknowledging the degradation we have caused on the Earth. We should rejoice and celebrate the good news that God has brought to all of creation. We commission each other to be good stewards of all that God has entrusted to us through the use of our gifts. As a result, our church's mission is enlarged. As Bishop James Jones so eloquently states, "The parish is the arena for the earthing of heaven locally. That is our local mission. World mission is the earthing of heaven globally."[3]

As represented in my childhood church experiences, we are well equipped to serve when there are both pastors and windows in worship.

Group questions

1. *What are worship experiences you have participated in that have given attention to both the spiritual and ecological fullness of God?*

2. *What are practices in your congregation that affirm and teach that everything belongs to God? How does this encourage holistic stewardship in your daily life?*

3. *How does your congregation affirm gifts that are used to care for creation?*

Toward ecological practice

In addition to considering how a congregation can include creation care in worship, there are creation stewardship practices to be considered as well. We could call this "greening the congregation." I suspect few churches would consider using the model of sustainability practiced in my childhood congregation. Our church was only heated on Sundays in the winter, and there was no air conditioner for those hot summer days – just the big windows to open. There were a few electric light globes hanging from the ceiling, but there was no running water – we had outhouses. Any meetings that needed to be held during the week – which were few – were held in homes.

Earlier chapters outlined the reasons why creation care is important from a faith perspective, major ecological issues existing today, and some practical responses we can have. The call is for each Christian to be grounded in caring for creation and to act accordingly. While individual actions are critical in restoring health in the environment, it is also important for the congregation to work together at discerning how to address ecological issues corporately. The following ideas are a good starting point for a congregation to explore and adapt in an effort to improve its stewardship of the environment.

> *The call is for each Christian to be grounded in caring for creation and to act accordingly.*

Create a creation care statement

This statement should serve as part of the mission and vision for the church. Stewardship of creation for congregations should grow out of a careful study of the Scriptures and discernment by members. As the church gains a renewed commitment to God's command to "till and keep" the Earth, a vision for creation care will emerge as part of its' purpose. The vision for this role should be integrated within the framework of the congregation's mission and core values. It is best if this vision and commitment is not just one more activity patched on to an already full agenda, but rather seen as an integral part of fulfilling the calling of the church.

Form a creation care leadership group

A good way to insure the ongoing integration of a creation care vision into the life of your congregation is to call out people with a passion and the gifts for guiding the church in this task. The group will function best if it can come alongside other activities of the church and gently encourage a fresh look at how caring for creation is part of each action. It is important the creation care group does not "radicalize" its activities, which is a sure way to drive others away from this important commitment. There are many ecological issues to address and various ways to work toward achieving the changes; therefore, acts of grace and wisdom from the group will most encourage the congregation.

The creation care group can provide leadership in many of the activities in this list. They may regularly ask church members to contribute new ideas and share actions that currently demonstrate the commitment to being earth keepers. The group can also assist in designing worship and education experiences that will guide people of all ages in your congregation to grow in love for caring for all that God loves.

Define the ecological setting of your church building

Just as individuals are encouraged by reflecting on place (Chapter 2), so the congregation can be inspired by learning to appreciate the place where the church is located. Property owned by congregations across the continent represents a great diversity of ecosystems. When members of the congregation begin to see their place of worship as more than a building and a parking lot, great strides can be taken. The following are several questions that can start the process of understanding our ecological place.

1. What kinds of soil are found on the church property?

2. What is the name of the watershed in which the church is located?

3. How many different kinds of plants and animals live on the church property?

4. How does the sun move across the sky in relation to the building's design?

5. Where do the prevailing winds come from?

6. What is the annual precipitation in your area?

Answering these and other questions help define location and begin to inform us on how we should best care for this place. Knowing this information – unique to each location – will also help members of your congregation understand your local community and how the church can model effective creation care stewardship.

Conduct an energy audit

Energy consumption in North America represents one of the greatest disparities in relation to our brothers and sisters around the world. For many years we have had access to cheap fossil fuel. In fact, our fuel prices are still cheap in 2008, relative to what it costs others on the globe. Therefore, we haven't been careful in controlling and limiting our rates of consumption. Even in our churches, we often let energy slip through our fingers.

A critical starting point to address this issue is an energy audit related to church functions. A good place to start is evaluating the energy uses in the church building and establishing a historical baseline of heating, cooling, and lighting costs. Next, check the efficiency of the equipment and determine changes that would reduce energy use. It is also important to make sure there are good seals on all doors and windows. Determine how much insulation is in the building, and consider how to add more to form a tighter building. Many times there are members in the congregation who have good skills in conducting this kind of assessment. A helpful guide has been developed by the U.S. Environmental Protection Agency called *Putting Energy into Stewardship: Energy Star for Congregations.*[4] This and other information can be downloaded from the Internet.

> *There are times we need to guard against making decisions based on frugality rather than on what is best for long-term ecological*

Careful assessments followed by improving the energy efficiency of the building can readily pay back the cost of the changes. This is wise stewardship. It may be hard to take the step of spending money on new, more efficient equipment and lighting, but because it saves on the use of fossil fuel (and pays for itself), it is a good decision. Other energy uses can also be evaluated such as transportation and food systems. In each case, this work can be viewed as part of the missional objectives of the church. Ironically, there are times we need to guard against making decisions based on frugality rather than on what is best for long-term ecological health.

Make a plan for reducing your ecological and carbon footprint

Other materials and resources should be assessed in addition to the energy demands of a congregation. Here are several areas to start.

1. What kind of paper is used in your copiers, printers, and for hand towels?

2. What alternatives, other than using paper, do you have for communicating with members and others?

3. What kind of cleaning supplies are used in the church? What is their environmental impact?

4. What happens to the waste generated by your congregation? How can this be reduced?

5. How might we make recycling a habit throughout the church?

6. What is our plan for using reusable items for food and beverages?

7. Where does our water come from and how much do we use?

8. Is our landscaping environmentally friendly?

9. How could we reduce storm water runoff from roofs and in parking areas?

An important time to weigh energy and resource uses carefully is when your congregation is preparing to build an addition or new building. The ideas on building green in Chapter 9 can also be applied to churches. Careful and wise planning in building designs can reduce energy and resource use and release funds for other ministries. It is also good to ask the question, "What are the alternatives to building that would still enable us to be effective in doing kingdom work?"

Design worship and meeting spaces to include visuals of creation

God placed us on the Earth as part of a beautiful creation. As we have noted earlier, having contact visually and physically with nature can enhance worship. I grieve when I enter meeting places that have few windows – it seems God's beloved creation has been blocked out of the sanctuary, which results in further alienation and potentially less care for what God loves.

Look at your place of meeting. Does it invite people in through an attractive design? Does the worship space provide views to the landscape surrounding the church? Are visuals used during the worship services that remind us of the Creator and our role as stewards of the Earth? The visual aspects of the church – both indoors and outdoors – are another area in which the creation care group can do some research and give insight.

Practice ecojustice in your community

Part of the local congregation's missional outreach in the community could be to participate in activities that promote ecojustice. Here are several ideas to get you started. Your list should be relevant to your location and should aid in bringing health to your community.

1. Clean up a stream, park, or roadside.

2. Assist in community garden programs – or even start one on church property.

3. Develop a Vacation Bible School based on creation care and invite the community.

4. Become involved with environmental groups in your area.

5. Promote good public transportation for everyone.

The church is more attractive to the world when we participate in these kinds of actions. Many times people who are passionate about the environment think Christians do not care about these issues. Working alongside people in environmental work is a wonderful opportunity for living and sharing the good news of Christ.

> *Working alongside people in environmental work is a wonderful opportunity for living and sharing the good news.*

Confess, restore, and celebrate

Including a time of confession for harm done to the Earth in our worship services is an important act in freeing us to practice creation care well. Confession is essential in overcoming our arrogance and self-centeredness. God desires restoration of all things (Colossians 1:20); therefore, restoration is also an act of worship. The mission of our churches should include restoration of hearts, people, and all of creation. Finally, we should celebrate God's goodness. This has been demonstrated to us in creation and in the incarnation. Christ's death and resurrection complete God's pattern for wholeness. We can rejoice with the psalmist by saying,

> "O sing to the Lord a new song;
>   sing to the Lord, all the earth.
> Sing to the Lord, bless his name;
>   tell of his salvation from day to day.
> Declare his glory among the nations,
>   his marvelous works among all the peoples." (Psalm 96:1-3)

Prayer:

Oh God, thank you for drawing us into relationship with other believers.

We praise you for the ways you declare yourself to us.

We seek you and desire to worship you in our gatherings.

May we do this with heart and mind and soul.

We confess that our buildings and ways of functioning

as a church deplete the Earth's resources.

Fill us with your wisdom so we can be better stewards in our congregations.

Grant us insights into how we can be missional as faithful keepers of creation.

May you receive the full glory.

Amen.

End questions

1. Which of the eight listed practices resonate with your congregation's mission and way of functioning? Why? Which would your congregation find most difficult to pursue? Why?

2. List additional ways in which your congregation can be better stewards of all the gifts and resources God has given.

3. What are creation care activities that could be achieved in your local area by working alongside people from the community who care about the environment?

An activity to do at home:

Take a walk (physically or virtually) around your congregation's property and building. Make notes on what impresses you about this place of worship. What are ways you could assist in reducing your church's ecological footprint?

Sources

1   Bishop James Jones, *Jesus and the Earth* (London: Society for Promoting Christian Knowledge, 2003), 92.

2   "Article 21: Christian Stewardship," in *Confession of Faith in a Mennonite Perspective* (Scottdale, Pa.: Herald Press, 1995), 77.

3   Bishop James Jones, *Jesus and the Earth* (London: Society for Promoting Christian Knowledge, 2003), 92.

4   http://www.energystar.gov/ia/business/small_business/congregations.pdf (page no longer exists).

12

Church-wide Creation Care

"The prophetic imagination knows that the real world is the one that has its beginning and dynamic in the promising speech of God and that this is true even in a world where kings have tried to banish all speech but their own. The task of prophetic imagination and ministry is to bring to public expression those very hopes and yearnings that have been denied so long and suppressed so deeply that we no longer know they are there." — Walter Brueggemann[1]

Purveyors of hope

Christians have a wonderful opportunity to be purveyors of hope in the midst of environmental degradation and the impact of climate change. The joining of believers in community forms the local church – creating a dynamic for ministering hope and being agents of change. There is a strength and synergy that arises out of worship, study of the Scriptures, healthy relationships, and discerning and engaging issues. There is both joy and struggle in the church as we faithfully embark on mission. In the letter to the Roman church, Paul writes about this tension.

> "Therefore, since we are justified by faith, we have peace with God through our Lord Jesus Christ, through whom we have obtained access to this grace in which we stand; and we boast in our hope of sharing the glory of God. And not only that, but we also boast in our sufferings, knowing that suffering produces endurance, and endurance produces character, and character produces hope, and hope does not disappoint us, because God's love has been poured into our hearts through the Holy Spirit that has been given to us." (Romans 5:1-5)

What a great motivational reminder to hear the words, "and hope does not disappoint us"! We carry a message of good news and reconciliation the world is eager to experience. Local churches join with other local churches to carry out in even greater ways the messages of hope for all of creation. This happens through denominational agencies, schools, camps, and parachurch organizations. There are amazing outcomes that result from people with many gifts and resources seeking common missional purposes. At the same time, it is important to listen to the prophetic voice of people like Wendell Berry, who in a lecture to a seminary audience states the adverse effect that the church has had on creation.

> Christian organizations, to this day, remain largely indifferent to the rape and plunder of the world and of its traditional cultures. It is hardly too much to say that most Christian organizations are as happily indifferent to the ecological, cultural, and religious implications of industrial economics as are most industrial organizations. The certified Christian seems just as likely as anyone else to join the military-industrial conspiracy to murder Creation.[2]

We are either unaware or have become desensitized to the impacts our local and church-wide activities have on God's creation.

I don't know about you, but this quote makes me squirm. I do not want it to be true – and yet I have enough experience and exposure to know there are truths in the statement. We often carry on our tasks – with excellent purpose and good intent – by following practices that cause harm to creation. We have grown so accustomed to our North American modes of operation, feeling they are normal, that we are either unaware or have become desensitized to the impacts our local and church-wide activities have on God's creation. In a hope-filled and encouraging chapter, Paul writes to the Corinthian church about what happens when there is not a focus on the message of Christ: "In their case the god of this world has blinded the minds of the unbelievers, to keep them from seeing the light of the gospel of the glory of Christ, who is the image of God." (2 Corinthians 4:4)

While Paul is referring to the blinding of people who do not follow Christ, we know that the gods of this world can also blind believers. The message God has entrusted to us is whole and complete. It is a salve – or salvation – that brings restoration to souls and all of creation. The good news Paul writes about is amazingly comprehensive. Repeatedly in 2 Corinthians 4, Paul exhorts us with the hopeful message, "Do not lose heart." He states we have a "treasure in clay jars," which brings God's extraordinary power into play in reconciling all things – which I believe includes caring for creation as we carry out the work of the church locally and globally. Through God's wisdom and Spirit, we can respond with a fresh sense of courage and determination. This will require finding new paradigms, new modes of operation, and a commitment to holistic, missional stewardship.

Through God's wisdom and Spirit, we can respond with a fresh sense of courage and determination.

The new paradigm is really an old one. It is what Jesus taught in the Sermon on the Mount (Matthew 5-7), the Sermon on the Plain (Luke 6), and throughout the Gospels. Christ's message is compelling, simple, thought-provoking, realistic, ambitious, and attractive – yet hard. The following statement articulates the application of Christ's precepts.

> "We believe that the church is called to live now according to the model of the future reign of God. Thus, we are given a foretaste of the kingdom that God will one day establish in full. The church is to be a spiritual, social, and economic reality, demonstrating now the justice, righteousness, love, and peace of the age to come."[3]

The commentary on this belief statement emphasizes that the church is to take seriously the paradigm – or framework – of Christ's teaching as our mode of operation today.

> "The church is called to live now under the rule of God as a witness to the reign of God. Our life together now is to be patterned after our life together in the age to come. This means that the reign of God is relevant to this world, and the ethics of God's rule should not be postponed to some future time."[4]

A consequence of being a church that practices justice, righteousness, love, and peace is the proclamation of hope to all of creation. When these patterns are infused into all of the church, we will experience renewal spiritually, socially, and economically. This renewal will surprise us with more hope! If we, as a church, were to live faithfully according to these teachings, the Wendell Berry quote would lose meaning – and we would no longer squirm under its indictment.

We – the church – have Christ's hope to offer to all of creation.

Hope is longed for today. Many people recognize the great dilemma of climate change and fear its consequences. This is pervasive in all aspects and parts of the world. And the need for hope is just as pervasive. Paul's letter to the Romans clearly states this.

> For the creation waits with eager longing for the revealing of the children of God; for the creation was subjected to futility, not of its own will but by the will of the one who subjected it, in hope that the creation itself will be set free from its bondage to decay and will obtain the freedom of the glory of the children of God. We know that the whole creation has been groaning in labor pains until now; and not only the creation, but we ourselves, who have the first fruits of the Spirit, groan inwardly while we wait for adoption, the redemption of our bodies." (Romans 8:19-23)

All creation is groaning. Redemption has come through Christ. While there is an anticipated completion of reconciliation and renewal, salvation is now present! Christ did not teach us to live carelessly on this Earth because there is a future of "making all things new." Regeneration is available now. It occurs within the human being – and it is a sustaining principle within creation. Even as creation groans, it is being renewed. We are called to be cocreators, coreconcilers, and cosustainers with Christ. And we – the church – have Christ's hope to offer to all of creation.

Group questions

1. How have you experienced the message of hope from Romans 5? In what ways might this passage be extended to all of creation?

2. What are your experiences that support the concepts in the Wendell Berry quote? What are observations you have made about the church that disagree with his statements?

3. In what ways do you see the belief statement on page 159 being lived out in church-wide activities? What are additional ways you might like to see this concept practiced?

The church in action

"The notion that religions might be influential enough to help shift whole societies onto new paths may seem fanciful. Religions typically lack armies, diplomatic prowess, control of legislatures, or other conventional sources of power that can shape a country's fundamental direction. But religious influence, often subtle and underestimated, can be astonishingly powerful." – Gary Gardener[5]

Jesus lived out hope in his daily life – and he challenged the powers of the time through new ways of thinking and acting. Time and again he confronted the hopelessness in society that various oppressors created. The impossible became possible in his new kingdom view. Christ modeled for his followers – and the church today – the way of regeneration and renewal. After healing a woman who had been crippled for 18 years, he described the kingdom of God in this way: "It is like a mustard seed that someone took and sowed in the garden; it grew and became a tree, and the birds of the air made nests in its branches." (Luke 13:19)

That which appears impossible – such as cleaning up a polluted river or changing our level of energy consumption – can be regenerated like the growth of a tiny seed into a large plant. Jesus proclaims that the church is the change agent for the world.

Taking what we have learned from the various chapters in this study can help us as a church to practice acts of creation care in ways that will plant seeds of hope in a world filled with despair. The following are brief descriptions of potential actions for us to practice in church-wide settings.

> *Jesus proclaims that the church is the change agent for the world.*

Integrating mission

The church is blessed with many great agencies that carry out various aspects of the mission of healing and hope. The commitment to creation care and holistic stewardship of all God has entrusted to us should be integrated into the process of fulfilling the primary mission of each agency. Each agency conducts work in local settings around the globe. Becoming familiar with the unique aspects of regional ecosystems can better inform how tasks should be pursued. Sensitivity to environmental impacts should be considered in each setting. It is best if we work out of the strengths found within the local setting rather than trying to impose processes that worked in some other location.

Demonstrating care for the Earth is a great way of witnessing. It is attractive when people observe us loving what God loves. Our love for all things – especially the least of these – communicates our depth of compassion, which is then quickly extended to those watching us. If our mission activity is negatively impacting the Earth, people will wonder about the consistency of our message.

Reducing agency footprints

The work of each church-wide agency leaves an ecological and carbon footprint. While there will always be some kind of impact on the environment caused by our actions, it is important we carefully plan for ways of reducing this impact. Each agency can benefit from having a green champion – someone who has a passion for caring for creation and consistently nudges the organization to implement the best practices. Establishing a creation care committee within the agency will aid in the assessment of the impact of current practices and assist in selecting improved ways of functioning.

> *Greater energy efficiency in buildings should be a priority.*

Some of the greatest impacts come from transportation and buildings. As we noted earlier, the burning of fossil fuel to power cars and planes produces large amounts of carbon dioxide, which leads to global warming. Reducing the amount of travel and choosing more efficient modes can reduce this impact. Well-designed green buildings help people function more effectively – and will reduce the ecological footprint. Greater energy efficiency in buildings should be a priority. Working carefully on both of these areas will free up more resources for achieving the purpose of the agency and will demonstrate a commitment to global justice.

Greening church-wide work

The church is a relational body. It is essential we interact with each other in ways that cultivate good communication and deepen our corporate commitment to the kingdom of God. Because building relationships is highly valued in our church culture, we gather people together for meetings, conferences, and assemblies. A resulting dilemma is the large expenditures of energy and resources needed to achieve these gatherings. The church, like much of North American society, is dependent on cheap fossil fuels to make meetings possible. It is relatively easy for us to travel to these gatherings with the energy from gas and oil. We like the places where we gather to be comfortable, the food to be good and quickly provided (with no clean up necessary!), the service to be excellent, all the current technologies to be available – and to be affordable!

But have we counted all the costs when we review the budget for these events? The affordability is largely due to the extraction of resources from the Earth that cannot be replaced and low pay for those providing the services. In order to be better stewards of all of creation, we must seek new paradigms for building and maintaining relationships in the church. This presents a challenge for the church to work on with imagination and wisdom.

Calling forth creation care stewards

In the first chapters of Genesis, God called Adam and Eve to "till and keep" the Earth. As we have noted in this study, this is a call that is still important today – especially when we have a commitment to holistic stewardship. The act of calling people to specialized tasks is a responsibility of the church. We have often thought of "calling" in relationship to pastors and missionaries, but it could also be applied to many other vocations. We have reflected on current ecological needs throughout this study that require an expertise to address. As a missional church, we should call out people with gifts and talents who are able to guide us in becoming better keepers of the Earth.

> *The act of calling people to specialized tasks is a responsibility of the church.*

Our church schools, colleges, and seminaries are places that can fill a critical role in this call. Intentional learning experiences in these settings will equip people to be effective as leaders and change agents in the stewardship of God's creation. The "greening" of the curriculum, the learning experiences, and the facilities at these institutions will assist the graduates in integrating creation care into lifestyle and vocation.

Sharing the Good News

I am thankful for the fullness of the Good News of Jesus Christ and the dynamic message of healing and hope found in the kingdom of God. Our church-wide agencies represent this message so well. Here are a few examples. The tagline of Mennonite Mission Network, "Together sharing all of Christ with all of creation," portrays the good news message with vibrancy. The many programs of Mennonite

Central Committee, set up to minister to "the least of these," are faithful responses to the Matthew 25 parable. The commitment of MMA to provide holistic stewardship opportunities – including environmental stewardship – is a proclaiming of the good news. Mennonite Economic Development Associates sponsor activities to alleviate poverty through green and sustainable practices.

Our God is a relational God. From the beginning, God set up a relationship with all of creation. Human beings have been given a special place and set of responsibilities in relation to God and all of creation. Salvation and reconciliation through Christ is a message that brings restoration. The Spirit comforts and nourishes us in the journey of following Christ. Caring for creation is an act of loving what God loves. The intersection of all these relationships brings delight and glory to God. May we be found faithful.

Prayer:

Oh God, our Creator, Redeemer, and Sustainer,

Praise be to your name!

You are majestic and lovely beyond words.

You formed us from the dust to be keepers of creation.

You established your kingdom through Jesus Christ.

You have called us by name.

You have anointed us with gifts to enhance your kingdom.

As members of your church, we give you thanks.

Grant to us wisdom and strength to be found faithful.

May you receive all the glory!

Amen.

End questions

1. *What excites you about being a missional church with a commitment to holistic stewardship in the 21st century?*

2. *What are the ways you see the good news proclaimed through church-wide creation care activities and commitments?*

3. *How do you see the church's response to ecological degradation being like the parable of the mustard seed?*

An activity to do at home:

> *Select a church-wide agency. Reflect on the mission it seeks to fulfill. Write a letter of encouragement to the leadership expressing your support for the vision of the agency and the efforts they are making to be responsible stewards of the Earth.*

Sources

1   Walter Brueggemann, *The Prophetic Imagination* (Minneapolis: Fortress Press, 2001), 64-65.

2   Wendell Berry, "Christianity and the Survival of Creation" in *Sex, Economy, Freedom & Community* (New York: Pantheon Books, 1992), 94.

3   "Article 24: The Reign of God" in *Confession of Faith in a Mennonite Perspective* (Scottdale, PA: Herald Press, 1995), 89-90.

4   "Article 24: The Reign of God" in *Confession of Faith in a Mennonite Perspective* (Scottdale, PA: Herald Press, 1995), 90.

5   Gary Gardner, *Inspiring Progress: Religions' Contribution to Sustainable Development* (New York: W. W. Norton and Company, 2006), 41.

Final Word: The Other Side

Congratulations! By now, you have probably read through the book and wrestled, personally, with some of the important questions it raises. That good step forward should not be discounted even though it may seem small to you. Many people never even venture this far.

The real question is what you do from here. The book and its discussion questions should have given you some clues, but here are a couple of other thoughts to get your thinking started.

- Change your home incandescent light bulbs to new energy-efficient bulbs.

- Plant a garden or buy food from a local farmer's market.

- Alter your usual transportation habits to take fewer trips, participate in a car pool, walk, or ride a bike.

- Find new recipes that do not rely as heavily on animal protein.

- Consider a higher-mileage vehicle for your next purchase.

The important point to remember is that there are small steps each of us can take to make a positive contribution for the environment. The hard part is to remember to think this way and to actually do something. We urge you to identify at least one small thing you can do beginning this week.

Bringing stewardship to life

If you enjoyed *Creation Care: Keeepers of the Earth*, you will want to consider the other books in MMA's *Living Stewardship* study series, including **Time Warped: First Century Time Stewardship for 21ˢᵗ Century Living.**

In *Time Warped*, Steve Ganger, MMA's director of stewardship education, provides:

- Twelve flexible, interactive lessons on how to "do less" yet create a more fufilling relationship with God.

- Practical Scriptural applications that ground each lesson in God's Word.

- Personal time chart and planning documents that help you take immediate action.

- Helpful group discussion questions that encourage deep, personal reflection.

- Encouragement and ideas that will motivate you to make lasting life changes now!

MMA's *Living Stewardship* study series examines holistic stewardship from the inside out. Each book deals with one area of stewardship – but in a holistic way.

You will think about stewardship in new ways as you work through these titles – and more books are in the planning stages now! Visit MMA-online (www.mma-online.org) to learn more about holistic stewardship.

Living Stewardship books and other educational resources, are available in the MMA Bookstore (www.bookstore.mma-online.org) or call (800) 348-7468, Ext. 269.

Bringing stewardship to life

If you enjoyed *Creation Care: Keeepers of the Earth*, you will want to consider the other books in MMA's *Living Stewardship* study series, including **Money Mania: Mastering the Allure of Excess.**

In *Money Mania*, Mark L. Vincent, a consultant with Design for Ministry, provides:

- Twelve flexible, interactive lessons that take readers beyond budgeting to visit various intersections in life where money plays a significant role.

- Practical Scriptural applications that ground each lesson in God's Word.

- Pointers to help you be earnest about your faith and organize your finances in ways that honor God.

- Helpful group discussion questions that encourage deep, personal reflection.

- Encouragement and ideas that will motivate you to make lasting life changes now!

MMA's *Living Stewardship* study series examines holistic stewardship from the inside out. Each book deals with one area of stewardship – but in a holistic way.

You will think about stewardship in new ways as you work through these titles – and more books are in the planning stages now! Visit MMA-online (www.mma-online.org) to learn more about holistic stewardship.

Living Stewardship books and other educational resources, are available in the MMA Bookstore (www.bookstore.mma-online.org) or call (800) 348-7468, Ext. 269.

Bringing stewardship to life

If you enjoyed *Creation Care: Keeepers of the Earth*, you will want to consider the other books in MMA's *Living Stewardship* study series, including **Talent Show: Your Faith in Full Color.**

In *Talent Show*, Bob Lichty, MMA's director of training, provides:

- Twelve flexible, interactive lessons on how to discover and use your gifts, talents, and passions in service to God, your community, and your family.

- Practical scriptural applications that ground each lesson in God's Word.

- Personal spiritual gifts inventory.

- Helpful group discussion questions that encourage deep, personal reflection.

- Encouragement and ideas that will motivate you to make lasting life changes now!

MMA's *Living Stewardship* study series examines holistic stewardship from the inside out. Each book deals with one area of stewardship – but in a holistic way.

You will think about stewardship in new ways as you work through these titles – and more books are in the planning stages now! Visit MMA-online (www.mma-online.org) to learn more about holistic stewardship.

Living Stewardship books and other educational resources, are available in the MMA Bookstore (www.bookstore.mma-online.org) or call (800) 348-7468, Ext. 269.

Bringing stewardship to life

If you enjoyed *Creation Care: Keeepers of the Earth*, you will want to consider the other books in MMA's *Living Stewardship* study series, including **Body Talk: Speaking the Words of Health.**

In *Body Talk*, Ingrid Friesen Moser, MMA's stewardship of health manager, provides:

- Twelve flexible, interactive lessons, that take readers beyond exercise, nutrition, and stress reduction to visit the various intersections in our lives where health, individual and community, play a significant role.

- Practical Scriptural lessons grounded in God's Word.

- Pointers to help you be earnest about your faith and about your health, individually and collectively.

- Helpful group discussion questions that encourage reflection.

- Encouragement and ideas that will motivate you to make last life changes now and going forward!

MMA's *Living Stewardship* study series examines holistic stewardship from the inside out. Each book deals with one area of stewardship — but in a holistic way.

You will think about stewardship in new ways as you work through these titles — and more books are in the planning stages now! Visit MMA-online (www.mma-online.org) to learn more about holistic stewardship.

Living Stewardship books and other educational resources, are available in the MMA Bookstore (www.bookstore.mma-online.org) or call (800) 348-7468, Ext. 269.